Met Man Pete Goes South

TROPICS
90°F/32°C

ROARING FORTIES
JUST A SLIGHT SWELL

SCREAMING FIFTIES
THE OCCASIONAL GALE

SCOTIA SEA
MIND THAT BERG, WHERE IS SIGNY ISLAND?

SIGNY ISLAND
WHAT'S ALL THIS WHITE STUFF?

Met Man Pete Goes South

Peter Richards

STOCKWELL
PUBLISHERS SINCE 1898

Published in 2022 by
Peter Richards
in association with
Arthur H Stockwell Ltd
West Wing Studios
Unit 166, The Mall
Luton, Bedfordshire
ahstockwell.co.uk

British Library Cataloguing-in-Publication Data:
A catalogue record for this book is
available from the British Library.
ISBN 9780722351369

Author's Note: This book was based on diaries I kept
at the time and I am so glad that the slaughter of
whales and seals is no longer acceptable. I hope I live
to see a worldwide ban on the killing of whales

Contents

To the five who lost their lives in the three years I was with FIDS.

Introduction

All my working life I have been a meteorologist. In 1957 I was working in the main forecasting room of the Met Office at Dunstable, at a time when we had many meteorologists working in all parts of the world. Many had worked as meteorologists during the Second World War, and it was exciting to hear all about their different experiences under varying conditions in places such as Africa, the Far East and Middle East, as well as on islands in the Pacific and Indian Oceans. However, none of these stories fascinated or interested me as much as those of my watch supervisor, John Lancaster, who had spent two years working in South Georgia in the subantarctic. John lent me Niall Rankin's book *Antarctic Isle*, which is about life on a small boat around the bays of South Georgia. It was captivating to read about the bird life that teemed around the coasts – albatrosses, penguins, and petrels – as well as the seals and whales. If any book could hook you on South Georgia, this book certainly could!

The Met Office produced a monthly *Met Magazine*, and on the back cover jobs were often advertised for meteorologists in Nigeria, the Gold Coast, Mauritius, Ceylon, the Seychelles, Gan, and many other exotic places; and in early 1957 there was a request for volunteers for South Georgia. Immediately, I applied for the post and was then called upon to attend an interview at the Crown Agents in London. There I was questioned about my interests, hobbies, and how I got on with others – nothing about meteorology! They must have been impressed with my replies as I heard that I had passed the interview, but I would have to go for a thorough medical examination, especially as I had failed my National Service Medical (Grade 4) because I had a perforated left eardrum.

Anyway, after a general medical I went to see a Harley Street specialist, and both passed me as fit for the job. In the meantime, unfortunately, another applicant was accepted for the South Georgia post. However, they offered me an even more exciting opportunity. How would I like a three-year engagement in the Antarctic with the Falkland Islands Dependencies Survey working at one of their Antarctic bases, leaving 1 October and spending two winters and three summers in the Antarctic, plus six months travelling there and back and leave at the end of the engagement? You can guess my answer. It wouldn't be a tropical island for me – more like life in the freezer!

I was advised to get an up-to-date passport and various vaccinations, and to visit a dentist because there would be no opportunity for any dental treatment for almost three years. They also told me to get any fillings 'insulated' to ensure permanency under Antarctic conditions.

I was to be a senior meteorological assistant; my agreement was as follows:

Quarters: They would provide quarters, subsistence, and routine canteen stores (including a ration of cigarettes, tobacco, beer and spirits) free of charge while I served in the Antarctic.

Passage: I would get free outward and return passage by sea on an expedition ship as a member of the supernumerary crew.

Personal: My personal gear had to be limited. The minimum requirement suggested was one suit, coat and sports jacket, a pair of flannel trousers, shorts and slippers, plus underclothing. We would leave our gear in Port Stanley, where they would issue us with polar clothing.

Pay: My salary was set at £440 a year.

On 19 September I received a letter directing me to report to the master on board RRS *Shackleton* by noon on Monday 30 September for documentation and the allotment of cabin accommodation. I would find the ship in Berth 37 at the old docks in Southampton. The ship would sail at approximately 3 p.m. on Tuesday 1 October; relatives would be welcome to see the departure.

In my five years working at Dunstable, I had built up many friends, and two of them, Tom and Rita Smith, arranged a farewell party at their house on Friday 27 September for myself and many of my workmates, including meteorologists, teleprinter operators and wireless operators. Tom worked in the wireless department while his wife, Rita, worked in the teleprinter room. We consumed a lot of booze and danced to records which were popular at the time, such as trad jazz and skiffle played by Chris Barber and Lonnie Donegan. To say farewell to my friends, I composed a calypso-type song to the tune of 'Rum and Coca-Cola':

> *Welcome all of you, my friends—*
> *Some of you work with plotting pens;*
> *Others bash out yards of tape*
> *Or spheric locations are your fate.*
>
> *Now, Alan here is going to stay*
> *And try to live on TP's pay*
> *For Rhodesia is now out—*
> *And New Zealand, I've no doubt.*
>
> *To Tom and Rita, for this do*
> *I would like to say thank you.*
> *Now I'm off to foreign parts,*
> *Far away from plotting charts.*

Alan never made it to New Zealand or Rhodesia, but he did go to Hong Kong and joined the Hong Kong police.

The next day, after the party, Alan, Tom, Rita, and I went to the Festival Hall in London to see a traditional jazz concert by Jack Teagarden and His All-Stars. Then it was home to Sidlesham in Sussex with my parents for two days to say farewell, before they took me down to Southampton to get my first glimpse of the RRS *Shackleton*, on which I was to sail to the Antarctic.

On arrival at the docks, they told us that the *Shackleton* was right down at the end of the old docks at Berth 37. Passing all the large cargo ships towering above us, which were loading and unloading, we headed for Berth 37, but all we could see were two masts sticking up above the edge of the quay. The *Shackleton* weighed about 1,000 tons – about the same weight as the *QE2* rudder. I reported on board to the captains and officers; then for the first time I met some of the other scientists heading south with me. For the outward passage, I was to share a cabin with a biologist, Fergus O'Gorman, who was to study the southern fur seal, which had become almost extinct due to earlier sealing in the Graham Land Peninsula. He was to spend his first summer on Powell Island, in the South Orkneys, and Deception Island, in the South Shetlands, and then he would return north to Montevideo in Uruguay to study there during the Antarctic winter before returning south for his second year to Signy Island in the South Orkneys. Alongside my other travelling companions, I was signed on as a supernumerary, which meant that we were paid something like one shilling a month and in return we had to keep the ship in shape, scrubbing down decks, painting, and helping with the loading and unloading of stores at all the places we visited. I think it was also a way of getting around insurance and regulations for when there wasn't a doctor on board. Anyway, after my first welcome aboard it was home for the final farewells.

The next day, 1 October 1957, it was once again off to Southampton, but this time not to return for nearly three years.

Part One: The Journey Down

(1 October 1957 to 23 November 1957)

Southampton to Montevideo

The *Shackleton* left Southampton at 3.15 p.m. on 1 October 1957. The weather was fine and the sea was smooth. As we passed down the Solent, we saw the *Ark Royal* aircraft carrier and *Mauritania* coming into port. We then entered the English Channel via the Needles.

On board, there were twenty-nine of us scientists altogether, commonly known as Fids (FIDS standing for Falkland Islands Dependencies Survey, whom we were working for). We were going down as meteorologists, surveyors, geologists, biologists, electronic engineers (doing ionospheric work), diesel mechanics, wireless operators, mountaineers-cum-general-assistants and a doctor. Most of us were going to be in the southern hemisphere for three years (two Antarctic winters and three summers) followed by six months of travelling and holiday. However, some people were only visiting the southern hemisphere for the summer; and the Doctor was only going to be in Hope Bay for a year.

The meteorologists took it in turns to do three-hourly weather reports from the ship on the journey down. We carried out these reports from the bridge, where the appropriate instruments were kept. One could find a small thermometer screen located on the rail just outside the bridge; inside it, we kept a dry, wet and sea thermometer. We used a hand anemometer so that by trigonometry the wind speed could be calculated, as the speed and direction observed was the wind speed and direction relative to the ship's movement. We would obtain the sea temperature by lowering a canvas bucket on a rope into the sea, and then we would put the sea thermometer into the bucket once we had hauled it up again. Occasionally, a hand would appear from a lower deck as we were hauling the bucket up and the water would be tipped out, but the culprit was never caught! Once we had collected all the weather data, we entered it into a register in the International

Meteorological Code. The officer on watch would supply the ship's exact location and we would then enter this in front of the coded weather message. We would then take the message to the wireless operator so that he could transmit it to the nearest radio station, who would then pass it on all around the world.

We passed through the Bay of Biscay in fine weather with only a slight swell. We spotted several birds, and some of them landed on the ship. These included a wren, a meadow pipit, a chiffchaff and a skylark – all migrating to warmer parts. The seabirds that we spotted were gannets, skuas, and gulls of several different kinds.

On the third morning at sea, we had our first taste of work at sea as we spent the morning holystoning the decks. We spent the afternoon in a more leisurely fashion, sunbathing on the forward deck. This became the routine each day, and we soon got around to doing some painting as well as holystoning and washing down the paintwork. We saw porpoises when we were off the coast of Portugal as well as wheatears and meadow pipits, which must have been on their way to North Africa. Each day we took it in turns to serve meals from the kitchen and to clean out the mess in the lounge. Every Sunday morning, all the cabins were scrubbed out and generally cleaned up for the Captain's inspection, when he came around with his officers to make sure everything was shipshape and tidy.

After five days at sea a few of the crew had gone down with Asian flu, so several of us went up on to the bridge at various times and learned how to steer the ship with the aid of a magnetic compass. Then, from when we were just off the Canary Islands until when we nearly got to Montevideo, the Doctor and I took over as helmsmen at the wheel during the first officer's four-to-eight watch each morning and evening.

On this watch, the first officer used to go out on the wings with his sextant to check our position at sunrise and sunset, so often I found myself as the only person on the bridge. In the dark if we sighted another ship approaching, we were told that as long as both his and our navigation lights were the same (for example, a green light to green light or a red light to red light) then we were OK. However, if we showed different-coloured lights to the other ship, we were in danger of crossing the other ship's path or set on a collision course with them. In this scenario, we would have to act. Nevertheless, down in the middle of the South Atlantic Ocean we sighted very few other ships; it was only when we neared the Brazilian coast that we saw many ships.

It was amazing to be on the bridge and see every sunrise and sunset. At one time, all but two of the crew were confined to their bunks with the flu so us Fids ended up also working the engine room and the galley. The Doctor's steering was most amusing at times: he sometimes let the ship get thirty degrees off course before bringing her around again. The course he steered was more of a zigzag than anything else! From up in the bows, you could see the wake of the ship curving its way out behind the stern, first on one side then on the other.

When the crew recovered from Asian flu, they took over the jobs of scrubbing down decks and painting while we carried on doing their work. The funny thing was that none of us Fids caught the flu. The doctor thought that this might have been due to all the different injections we had had just before leaving, which could have made us immune to that strain of flu. At one time, we were going to call in at Dakar for fresh water, but we were not allowed to enter the port when the port authorities heard that we had Asian flu on board. Therefore we altered our course for Montevideo and used the water we had more sparingly, although we never got to the state where we had to ration it. It was also a good excuse for drinking more beer! The journey was a long-haul trip from Southampton to Montevideo – a four-week journey at sea.

Each of us, in turn, were invited to have lunch with the ship's captain, Norman Brown, and his officers in the wardroom. My turn came as we were passing the Canary Islands. I went with 'Tinker' Bell and spent a very enjoyable three hours with the officers chatting. The weather outside was perfect, and I enjoyed seeing the Canary Islands as we passed.

One of the favourite pastimes when we were not working was to lean over the bows of the ship and try to spot fish and porpoises swimming in the water just ahead of it. We saw a lot of flying fish in the tropics. At night several of them came on to the boat and were found on the deck the following morning. The odd one or two even came through an open porthole into the cabin.

Flying fish are food for many other types of fish and dolphins; their ability to fly is their main form of self-preservation, keeping out of the way of their enemies. We saw three different types of flying fish.

The first kind we saw, which was the kind that came on to the deck and through the portholes, averaged in size from a few inches to a foot in length (around 7–30 cm). Often, they were in shoals of up to a hundred or so, but it was not uncommon to see them in ones and twos. When they are in the water the most distinguishing thing about them is their bright-blue backs – they could often be seen swimming in the clear water just in front of the bows of the ship. Their bellies are a silver colour.

A feature of this type of flying fish is that their wingspan is one and a half times the distance from their nose to the tip of their tail. Their tail fins are unusual as the tail's bottom half is about twice the size of the top half; this appears to assist them when they use their tail to flip the water as they glide through the air to give them extra propulsion or to alter direction. To fly, the fish must build up speed quickly in the water and then, with a flick of its tail, leap into the air and expand its wings. They use their wings to glide through the air, and if the fish begins to lose momentum it gives a flick of its tail on the surface of the water, which enables it to continue gliding.

They glided any distance up to fifty yards or more. Sometimes, when they returned to the surface of the water, they would give a flick with their tail and would change the direction of their glide at almost right angles. They would sometimes do this two or three times before entering the water again, but each time they turned in the opposite direction. This was so they would continue to get further away from you in a zigzag path as opposed to going around in circles. When they had finished their glide and entered the water again, they would close their wings against their body before they touched the surface.

The second kind of flying fish was very similar to the first except they had another smaller set of wings set further back on their body.

The third kind was very unusual, and we only saw them on a few occasions south of the equator. They had a rather tubular-shaped body with a set of wings at either end of it. The longer set of wings, near their tail, was a reddish colour, and the shorter pair of wings, near their head, was a bluish colour. It appeared that as they propelled themselves into the air they shot water from their tails, so that when they first left the water there was a stream of water coming out behind them. Their flying ability didn't seem to match up to that of the other types of flying fish as they only managed to travel about five-to-ten yards (4.5–9 m) before entering the water again.

Alongside flying fish occasionally coming in through the portholes, the fact that I shared a cabin with a biologist meant that several specimens were brought by him into our cabin and were deposited in our washbasin.

For our amusement, we played football with cigarette tins on the afterdeck. At other times, when the ship was pitching in the oncoming sea, we would go up to the bows, and when they rose high with the sea we would jump into the air and then come crashing down, as the deck would have dropped in the swell.

When we reached warmer seas, we noted that the water became a deep blue in colour and that we could spot fish beneath its surface. For example, when we were south of the equator, just off the Brazilian coast, we saw Portuguese-men-of-war jellyfish. However, as we neared Uruguay's coast, the sea, many miles out from the coast, took on a brown colour from the waters of the River Plate.

At one time, the sea temperature rose to 28°C (82°F), while the air temperature was 30°C (87°F). It remained much the same as that all throughout the tropics. The weather nearly all the time was fine with only a slight swell. When at sea, we used to get out on deck and have a sing-song after the evening meal. Tony Richardson, with his guitar, was the main instigator behind this entertainment. He was accompanied at times by another guitar, a harmonica and a washboard.

We had film showings about every third evening in the mess after the dinner. These used to be quite a laugh as the electrician used to play different bits of the film over again or play them backwards at the request of the audience. One favourite was *The Gunfight at the OK Corral*, and the gunfight at the end of the film would sometimes be played in reverse with dead men jumping up and running backwards. Most of the films were five or more years old, and it seemed as if every other one had Jack Hawkins in it. In the end, every time he appeared in a film a big cheer went up from everyone.

When we were near the equator, we saw a lot of phosphorescence at night, which looked just like gun flashes. Off the Brazilian coast one evening we found a very colourful butterfly on the deck; it must have been carried out to sea by the offshore breeze. Also, while we were in this area, a Brazilian anteater landed on the upper deck for a short while.

When we had been at sea about three weeks, they told us that all people who were going to make weather observations would have to learn to send and receive Morse code, as we would be down south in the International Geophysical Year, which meant that we would have to send out weather reports from our bases every six hours, both day and night. Each base only had one wireless operator; therefore, as that person could not be expected to be on call every day and night, and as a useful backup in case they were taken ill, they organised a crash course for us in the mess each afternoon so that we could learn Morse code and practise tapping it out. But pity the poor operator at the receiving end when we first started to send messages!

We held a knockout darts contest during the trip down (I went out during the second round), with the final being played on the night before we entered Montevideo. This was a great party night with everyone in high spirits. Land at last after a month at sea! The Doctor, amongst other people, gave us a talk on various topics from the perhaps undesirable women we might meet in the port to frostbite in the Antarctic. Going up the River Plate, the Captain was on top form as he managed to turn the ship around in a full circle, with only the aid of a rather drunk third officer, in a channel that could not have been more than a mile wide, while other ships were going up and down it!

Next morning at ten thirty we docked in Montevideo.

Uruguay in the 1950s

Uruguay is situated in the continent of South America between the countries of Brazil in the north and north-east and Argentina to the west. Along the southern edge of Uruguay is the River Plate and to the south-east the Southern Atlantic Ocean. The River Uruguay flows along the boundary between Uruguay and Argentina. Montevideo is the capital and the main port of the country.

The people seemed to be divided into two classes: the very rich and the poor. There seemed to be very few middle-class people. The main occupation of the country was farming – agricultural farming along the southern coast and cattle farming in the rest of the country.

The crops grown in the south were wheat, maize, flax, sunflowers, rice and, to a smaller extent, beetroot and sugar cane. They also grew fruit such as mandarins, grapefruit and lemons in this area. Further north, they reared cattle, horses and sheep. Wool from the sheep was the main export of the country. There was also a fair growth of timber along the south coast, where you could see trees such as firs, eucalyptus, laurel and palms. They used fir and laurel trees quite extensively as windbreaks on the fruit farms.

On each side of the roads away from the town there were twenty to thirty yards (18–27 m) of wasteland left before the fields so that the cattle could be driven along by the cowboys on horseback (often with the aid of Alsatian dogs). The cowboys would wear cloaks and sombrero hats.

The roads were straighter than the British roads, but they were in very poor condition. Only the main roads in the towns were asphalt, cobble or concrete; the rest were just loose gravel with many holes and bumps. The architecture seemed to be more modern than the British designs for houses, bungalows and blocks of flats, and they made a lot of use of wrought ironwork, especially for gates.

The main sport was football, and on any space of waste ground children could be seen kicking a ball around. Other common sports were bike racing, basketball, horse racing and polo.

The main export of the country was wool, which constituted on average to forty per cent of Uruguay's exports. In the year 1950 this rose to sixty per cent, which was the highest that it had ever been. They had an income of 65–70 million dollars that year from wool alone. Meat and its by-products (such as hide and leather) made up another twenty per cent of Uruguay's exports.

A new house cost about £14,000, the latest new American car £4,000, a New Morris 8 £1,400, and an Anglia over £1,000.

The police looked like the American policeman, with truncheons and revolvers hanging from their belts. They had equipped the cars and motorbikes with sirens.

Montevideo

We spent much of the daytime sightseeing around the town and shopping, with of course a steak meal every day – lovely thick tender juicy steaks! We made two visits to the cinema – the films that the cinema was showing were nearly all American, with Spanish subtitles at the bottom. It was quite amusing as we were ahead of the rest of the audience and saw the jokes long before they did! The cinemas on average were much smaller than the British ones at the time, although they did have much more leg room between the seats and we did not have to get up to let people pass us. No smoking was permitted in the cinemas. The films we watched were *The Man Who Knew Too Much* with James Stewart and Doris Day, and *The Desert Fox* with James Mason.

Walking through the town on my second day there was a chap who came up to me and asked me if I could speak English. It turned out that he was English too! He was called Brian Leary and came from Kingston – he thought I was not a local because of my clothes and hair colour. We had quite a long chat. He had just arrived the day before us and was in an American ice show; he was lost and wanted to know where the information bureau was, and, as it so happened, I knew and was able to help him. He was busy rehearsing, so I was unable to meet him that evening, but I phoned him the next day with the help of one of the locals who got through to his hotel for me, but we did not meet again because he was too busy, and I did not know for certain when I would be leaving.

We arranged and went on two coach trips. One was an afternoon trip around the local area, which was quite interesting as we saw the new skyscraper hospital and the big new football ground. The second trip was very interesting and lasted from 7.30 a.m. to 7.30 p.m. (200–240 miles). We went by an inland route to Punta del Este via Pando.

On the way we had to stop for a big cycle race to go by – the police had stopped all the traffic and made it all pull in to the side of the road. Punta del Este was a kind of millionaire's holiday camp situated at the mouth of the River Plate on a peninsula with sea on three sides of it (it was one of the stops on the around-the-world-sailing trips).

The hotels and gambling casinos were massive places and of modern design. Trees surrounded the grounds of each house. The park was closed for holidays, so the place was quiet, but we had a good look around. The people who worked for the coach company laid on a big dinner for us at one of the hotels – five courses, wine, the lot! We then returned via the coastal route, visiting Piriápolis. The whole trip cost us about twenty-four shillings, which was a real bargain.

I fished over the side of the *Shackleton* twice while we were in Uruguay. On the first occasion I caught thirty-two fish and on the other occasion forty-three. The fish looked very similar to bass – the local name for them was *corvina negra*. The chief cook supplied us with fat and let us cook them in the galley in the evening, provided that we washed and cleaned up afterwards.

We also used to upset our Welshman, Mike Crockford (Taffy), by singing an English version of the Welsh national anthem.

> *Wales, Wales, bloody great fishes are Wales.*
> *They don't come in tins.*
> *They're not sardines.*
> *They're bloody great fishes are Wales.*

The last night we were nearly all broke, but we got together all the money we could muster and visited the Can-Can bar and had an amusing and slightly interesting evening there. 'Tinker' Bell got drunk and did cartwheels all along the main street! 'Fritz' got himself into a similar state, but he just curled up on the floor and went to sleep. Tinker took a trip back to the ship and came back with some cans of beer to keep us going a bit longer. He got by the harbour police by stopping and singing to them – they must have thought he was going around the bend! Don Hawks and I thought we would try to do the same, but the police stopped us and took the beer away – obviously we did not impress them with our singing. I bet they drank the beer later that night themselves! Still, we went back and joined the boys for a while until we did not have a peso between us; then we thought we had better leave.

The next morning, we anchored in the harbour away from the docks and took on board aviation fuel for the Beaver seaplanes in the Falklands. Then we sailed out of the harbour in the afternoon.

Montevideo to the Falklands

We left Montevideo on 1 November at 3 p.m. The first night at sea we had violent thunderstorms and lightning; the next day the weather was the roughest that we had experienced so far, with waves crashing over the bows. The whole ship shuddered at times as the bows came right out of the water and as the ship pitched in the heavy oncoming seas. One person said that if the ship's bows come out of the sea three times in a row and the ship shuddered, the ship would break its back. Thankfully, we never experienced three shuddering jars consecutively to find out. In these southern waters, many birds followed the ship, including albatrosses, giant petrels, great atlantic Shearwaters, Wilson petrels, silver-grey fulmars, and Cape pigeons. It was too rough to venture on deck too much, so no more sing-songs under the stars, but they continued in the FID'ary (our lounge). On the second night, after the storms, the engine fuel pump broke down, so we drifted for twelve hours until someone was able to fix it. Thank goodness the weather had moderated! However, it had turned foggy, and it was rather eerie drifting through the fog without the sound of the engines. When we were a day away from the Falklands, we saw our first whale. Finally, after six days at sea, we arrived on a windy day with sleet showers into Port Stanley Harbour on the Falklands.

6 November

We first glimpsed heavy seas crashing against the Falklands' barren rocky coast at 10.30 a.m. on 6 November. Many birds pursued the ship as we neared land, and some black-and-white Commerson's dolphins bobbed up and down in front of the bows. After squeezing up the narrow channel which led to the entrance of the harbour, we were greeted by a sleet shower as we docked in Port Stanley at 3.30 p.m.

Port Stanley

The view which greeted us to Port Stanley after passing through the narrows into the inner harbour was of the Two Sisters framed behind the Town and with Mount Tumbledown on the left and Mount Kent on the right, shrouded in cloud.

That evening a packed reception was held on board, to which the Governor came plus 113 other guests. Suffice to say, the ship was extremely crowded! The two main centres of gathering were the wardroom and the FID'ary. There I met Mr Canning, who was the Chief Meteorological Officer for the Falkland Islands and the Antarctic. Also, I had the opportunity to talk with Dick Smith, the senior assistant in the Falklands, whom I had first met in Dunstable – not forgetting Joan, Frieda, Rosemary and Penny from the FIDS office and Susan, the Governor's daughter, whom I also talked with. Afterwards I was invited back to Frieda and Eric's home, with the folk from the FIDS office and many others, including Jack Richardson (chief engineer) and Tom Flack (first officer). Just like me, Eric, Frieda's husband, at one time had worked in the Met Office before joining the FIDS. After his spell in the Antarctic, however, he remained in the Falklands working as the FIDS treasurer in the Stanley office. That evening, Eric excited us all by showing us many coloured slides he had taken at different times down at the Antarctic bases.

7 November

The next day, I explored Stanley in the morning, I then heard for certain that I was destined for Signy Island in the South Orkneys. In the afternoon, I visited the Met Office with Fritz and Robin Perry. There Dick showed me the new wind-speed recorder which I would have to install at Signy when I arrived there. That evening, we all got invited to a reception at the Governor's house, where I met up with

Brian Beck, who was also destined for Signy Island. Prior to joining the FIDS, Brian had worked at a sheep farm on Pebble Island. However, he had decided that he wanted a change, so the meteorological office at Stanley trained him in meteorological observing practices. I also saw the Mr Canning's daughter Maria, who was the centre of attention for quite a few of us boys – that was until her boyfriend showed up! Afterwards, about seven of us meteorology types traipsed off to Pat Canning's place for supper.

8 November

I whiled away the entire morning up at the meteorological office chatting with Dick before going up to the FIDS office in the afternoon, where I spoke to Joan about getting some calendars to send home. Then at 4 p.m. came our football match against the crew! We were at a distinct disadvantage as all the crew had a full kit, whereas we had to make do with shorts and plimsolls. Unfortunately, we lost 4–3. I played left back, and did I have some aches and pains after the match. I hadn't done so much running around for a long time! In the evening, we all trooped along to a dance held at the Town Hall, where the Governor was also in attendance.

9 November

I was back up at the FIDS office in the morning, and after lunch a crowd of us, including the second officer, Tom Woodfield, took the ship's motor boat across to where the wreck of the *Great Britain* lay in Sparrow Cove. Some of the lads went shooting and returned with four geese and three snipe; meanwhile the rest of us explored some of the gentoo and jackass penguin rookeries. Earlier in the day, after having stayed up all night drinking, 'Leckie', the ship's electrician, and Tom Richardson managed to shoot two Logger ducks before breakfast.

Therefore that evening we decided to have a barbecue at the beach. We started plucking the birds in the FID'ary – what a job it was! The Captain just so happened to be holding a cocktail party in the wardroom above, and he sent a word to us downstairs to keep all the doors shut as the smell kept drifting upstairs. Thankfully, the second cook came in and kindly gutted most of the birds for us. However, the smell was too strong for Taffy, who had to leave. Later he went and got a beer for us. Also, 'Lamps' said he would give us a hand,

and proceeded to half pluck and half skin one of the birds. He then took it outside and split it open with a knife from its tail to its throat before tipping its guts over the side of the boat. He then took the bird, dropped it in the toilet and pulled the chain. Finally, he presented the bird to us, asking us to give him another one to do – a request we declined before later dumping the bird he had already prepared.

His prize remark was 'You ought to shoot them in the moulting season, you know. Feathers would come off much easier then.'

As some of the birds were protected, we had to dispose of all remains of them, so I got a sack of shingle from the jetty and half emptied it, then put all the evidence in it before ditching it over the other side of the boat.

The completed birds were all wrapped in newspaper and were taken down to the beach, where some of the others, under Gordon's guidance, had got a big fire going made from driftwood. We cooked the ducks, geese and snipe for two hours, basting them with oil from the fat we had acquired from the galley, alongside roasting potatoes which we had stolen from the forward hold. The night was rather chilly, so we got very tired of waiting and waiting for the birds to cook, so we ate most of the roast potatoes and drank the beer. When finally we decided to eat the birds, we drew lots to decide who should have the biggest bird. It worked out that there was one bird between two. The first slice was lovely; however, further inside we discovered that the bird was not cooked at all and the meat was still blood red! In the end, very little was eaten, and those who came off the best were actually those with the smaller birds as the smaller birds had cooked better than the others. Most of the birds were thrown into the 'og' (a nickname for the sea used by the seamen and Fids) except for a few slices which had been cut off the outside. We all returned to the boat at about 1 a.m., dirty, tired and still hungry!

10 November

Several of us decided to go along to the customs officer's office and hire the 'Alert' local motor boat plus a boatman for the day to take us across again to Sparrow Cove, where we had been the day before. Some of the chaps climbed Mount Low – 671 feet (204 m) – while some of the others went birdwatching. Gordon and I decided to walk across to the further coast via a stone-run on the western side of Mount Low and visit a rockhopper penguin rookery overlooking Berkeley Sound. Their nests were on the tops of cliffs 150–200 feet (45–60 m) above sea level. We collected some eggs and returned to Sparrow Cove via the other side of Mount Low. We then got a fire going and cooked the eggs plus some other food which we had brought with us from the ship for our tea.

11 November

I went along to the stores to collect all my Antarctic kit – long johns, mukluks and many other things. After lunch I went for an interview with the Governor – we chatted about birds and photography. Later I found out that he talked on various subjects with all the other boys and seemed to be very knowledgeable on a wealth of topics. In the evening, John Green, secretary to the FIDS, came to visit me on the ship and told me that I was going to be the base leader at Signy Island. Well, I was very surprised! He told me to go up to the FIDS office the next day to see him. Later that evening, I went along to the Town Hall to the dance put on by all the men going south. Tony Richardson, Jim Franks, Alan Gill, and Duncan Boston got up onstage and sang their theme song, 'Oh, the Fairies'.

> Oh, the fairies—
> There's nothing so splendour as
> feminine gender.
> Oh, the fairies! Oh, the fairies!
> Oh, for the wing of a fairy queen!

Plus, other verses that I cannot remember, but it ended with a shout of

> And to hell with the Chatham to London Railway.
> Sleepers awake!

12 November

I went up to the FIDS office and got sworn in as base leader and magistrate for Signy Island, in front of the Governor. I then spent quite some time looking through the operational instructions and files. After lunch, I went up to the meteorological office to help pack up the new instruments and gear for Signy. In the evening, I met Richard Smith and Jim Shirtcliffe (a former Signy islander).

13 November

In the morning I was up at the meteorological office again, this time to work out a new correction card for the barometer which I was to take down to Signy with me. Then in the afternoon I was up in the FIDS office again to learn about codes and ciphers for secret messages. Frieda was my tutor for this, but she had very little time to spare as people kept on popping in for all sorts of things at the last minute. In the evening, I went along to a party at Joan Ward's. Also at the party there were Fergus, Norman, Eric and Frieda, Penny and Bill, Tom (second officer), Susan and Rosemary. When I got back to the ship I found some mail and birthday cards for me that had arrived in the RMS *Darwin*, which had docked at 9 p.m.

14 November

The last day in Stanley was spent visiting the Treasury to hand in the money I had not spent. Then, I went up to the FIDS office once more to try and pick up a bit more of the codes. There were two different kinds of secret codes. One dealt with the movement of naval ships because HMS *Protector* toured the Antarctic bases during the summer. This code was also to inform us if any VIPs, such as the Governor of the Falklands, were travelling around on the ships. It would also tell us about any foreign activity in the area – for example, if the Argentines were to set up a base on our island, or if there were any hostile activities. The other code was more for personal matters – for instance, if we wanted to pass messages to the FIDS office in Stanley in confidence on personal issues. I went to Joan's in the evening before we left, with Fritz and Alan Cameron, to play some new records so Fritz could tape-record them. We then returned to the ship just before we sailed at 10.30 p.m. All the people from the FIDS office were there to wave us off.

Southampton, Berth 37, RRS Shackleton.

Port Stanley.

Icebergs

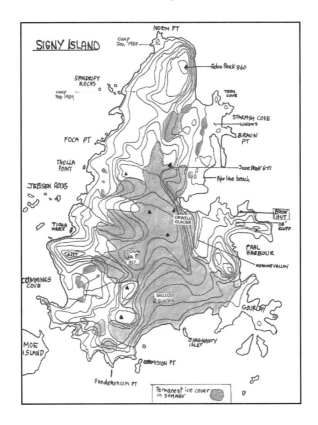

Base hut on Signy.

Lenticularis cloud over Sunshine Glacier on Coronation island.

Peter sawing out snow for the water supply.

Our home, Tonsberg House, Signy Island.

Robin Peak with Coronation Island behind, taken from Moraine Valley.

Cape Hanson and Sunshine Glacier on Coronation Island.

Giant petrel on nest.

A well-ringed skua. On the right leg is a blue-and-white band,   used in the IGY only on Signy; on the left leg, yellow/ red spiral rings for local tracking. This bird was a regular visitor to base and nested at top of the Stone Chute.

Tea break while ringing giant petrels on the west coast of Signy.
Left to right, Peter, Alan Sharman and Gordon Mallinson.

Doug Bridger and Peter's home, a two-man tent,
while surveying at North Point, Signy.

Alan ringing a giant-petrel chick on the west coast of Signy.

Signy wintering party, 1958. Photograph taken midwinter.
Back row: Gordon Mallinson, Alan Sharman and Peter Richards.
Front row: Brian Beck, Jim Stammers and George White.

The Falklands to Signy via South Georgia

The journey from the Falklands to South Georgia is usually accompanied by foul weather and rough seas. However, we were lucky as the sea was fairly smooth when we travelled, but the weather was colder. On our third day out we saw our first iceberg after we had crossed the Antarctic Convergence. South of the Convergence the sea temperature never gets above 4°C (39°F). The Convergence is quite a distinctive line all around the Antarctic continent, and when it is crossed the sea temperature dips by several degrees. On this leg of the journey we spotted many albatrosses and petrels of various species. I also made my first attempt at hair-cutting, which turned into a regular job once I got to Signy Island.

When we stopped first at King Edward Point, Government Station, we unloaded two bungalows in sections which we had brought down from Southampton. We had also brought four builders down from Port Stanley to do the erecting. In the bay behind King Edward Point was Grytviken, the Argentine whaling station. We saw several whales brought in by the catchers and hauled up on to the plan and cut up.

On the beach near Grytviken there was a dead seal which stinkers (giant petrels) were eating. If they were frightened away, they could not fly because they had eaten so much! Instead they just waddled over the beach into the sea and then flapped along the surface. The next day, all that they left of the seal was a few bones. That evening, we sailed up the coast to Stromness, the fuelling and repair station for the British whaling factory at Leith in the next bay. We visited Leith by barge to have a look around the factory and the slop chest (shop for the whaling station), where we could buy lovely thick Norwegian sweaters, skis, sweets and many other things. There were two factory ships belonging to Leith – the *Harvester* and the *Forester*, and twenty-five whale catchers which worked with them. There were about twenty men to each whale catcher. Twelve whale catchers worked with the *Harvester* and thirteen with the *Forester*. There were also a few other whale catchers which worked from the land station.

The whalers were mostly from Norway, Scotland and the north-east of England. Their pay ranged from £500 to £3,000 a year. Usually they were out there for eighteen months at a time – one Antarctic winter and two summers.

The whaling stations had everything they needed, which meant that they were totally self-sufficient. They even had, amongst other things, a blacksmith (to straighten and reshape harpoons), a carpenter, a hospital, a radio station (to keep in contact with their whale catchers), a shop (slop-chest) where you could buy all day-to-day items, such as clothing, soap, films, chocolate, etc., a cinema (Kino) which showed three different films a week in the summer, and a church.

Each whale was worth around £5,000, and almost no part of the whale was wasted. The people working at the land station could handle about six whales in eight hours. The smell around the factory was horrible. It was the smell of whales being cut up three to four days after being killed, but the men who worked there said you would grow accustomed to it after a while. When they cut up the whale, the sea turned red with blood for a hundred yards around the plan. Usually there were several hundred Cape pigeons on the water near the plan, feeding on all the waste washed out to sea from the factory, and usually quite a few stinkers (giant petrels) as well.

At the whaling stations, everything orientated around the plan, which was a large wooden area at the water's edge where the whalers hauled up the whales using steam-powered winches. As soon as they started to haul the whale out of the water on to the plan, a 'fencer' at the edge of the water made the first cuts into the outer skin and flesh. First, they cut off the blubber to be rendered down to oil. Then they sent the flesh, stomach and finally the bones all different ways to be cut up and processed. Once the whale was hauled out, the whalers could completely cut it up and send it away to the boilers to be processed in thirty to forty-five minutes, depending on the size of the whale.

The workers there were not allowed to drink any alcoholic drink at the factory. When our ship came in the workers tried to buy bottles of spirits. They offered £5 for a bottle of whisky and £1 for a can of beer. We were warned not to sell under any circumstance.

We left South Georgia at 6 a.m. on 20 November, travelling out along the east coast of the island. The scenery was very impressive with snow-covered mountains and glaciers coming down to the sea. We could see many penguins porpoising through the sea while the albatrosses and petrels filled the skies. Off the southern tip of the island we passed a whale catcher heading home to one of the factories with a whale tied, tail first, to the side. Once we had rounded the southern tip of the island, we headed south-west for the South Orkneys and Signy Island.

We saw many icebergs. The seas were very rough for the next couple of days. By midday on the 22nd we had seen our first pack ice. We entered the ice during the afternoon – everyone was busy with their cameras photographing. The ice slowed us down quite considerably as we nosed into it and broke it up to forge a pathway. Occasionally we had to go back and forth to break the ice. Some of the icebergs were over fifty feet (15 m) high and over a quarter of a mile in length. Pack ice is frozen sea. Where we were the sea ice most likely originated from the Weddell Sea, and had been carried north by the sea currents and winds. Icebergs are pieces of glaciers and sea ice from around the Antarctic continent which have broken off. We spotted many silver-grey fulmars, snow petrels, Cape pigeons, prions, and albatrosses. We sighted the South Orkneys at 4.30 p.m. We should have arrived at Signy Island that evening at about 11.30, but due to the pack ice between Powell Island and Coronation Island we did not arrive at the base until 6 a.m. on 23 November.

After breakfast on board the *Shackleton*, I went ashore to meet Cecil Scotland, the outgoing base leader, and the rest of the previous year's company. We were the first visitors for them since the previous April. Cecil spent the next two days showing me around and handing over the base to me and my compatriots. These two days were very frantic as we had to unload all the stores and supplies to last us for a year and take as much of it as possible inside the hut. Boxes, boxes everywhere! This meant that later we had the great job of sorting them all out and storing them away. The only things left outside were the drums of diesel fuel, for the generators, and sacks of coal.

At the end of the second day, the *Shackleton* left with all the previous year's company except for the two surveyors, who had the summer to finish off their final surveys of the island. This left Alan

Sharman, Brian Beck (meteorologists), Gordon Mallinson (wireless operator), Geoff Stride (diesel mechanic), Doug Bridger and Rob Sherman (surveyors), and me on the island. The *Shackleton* was bound for Powell Island to land a party there, including Cecil, to survey the island during the summer.

An Interlude: Early Discovery and Development of Signy Island

Signy Island was first discovered in 1821 by George Powell on a sealing expedition from England. There were not many seals on the island, which meant that it held little interest for sealers and was therefore only infrequently visited. Whaling first started on the island in the 1907 and 1908 summers with a floating factory plus two catchers. After 1908 there was little activity on the island until the early 1910s (prior to the First World War), when floating factory ships again visited Signy. In 1913, the *Tioga* whaling ship was blown aground in Port Jebsen in a gale and sunk. Part of the ship could still be seen protruding out of the sea in 1958. There was a shore station established in Borge Bay for four years in the early 1920s. Also, floating factory ships operated in the bay intermittently up until the 1930s.

The early whalers did not name many of the features on Signy Island. Petter Sorlle named the island Signy after his wife, Fru Signy Sorlle. The hut we lived in was named Tonsberg House after the company that operated the site in Borge Bay. Pipeline Beach and Pumphouse Lake on the island were used by the whalers to get fresh water supplies from the lakes when they were not frozen. The highest point on the island, Tioga Hill, was named after the whaling ship which sunk at Port Jebsen. Many signs of the whalers remained on the island – for example, the hut and pipes on Pipeline Beach.

On the site of our hut there was also the old whaling plan – a flat wooden surface for hauling up whales to be processed – and there were various bits of old winches and barges nearby. There was also the whalers' cemetery in Moraine Valley, where there were the graves of three whalers.

Signy Island was known as Base H, and it was called the 'banana belt base' by many because it lay further north than the other bases on the Graham Land Peninsula and the Antarctic continent. Signy is about sixty degrees south of the equator, which is roughly as far south of the equator as the Orkney Islands in Scotland are north of the equator. There the similarity ends, as the Scottish Orkneys are kept relatively mild by the Gulf Stream whereas the South Orkneys are cold because they are south of the Antarctic Convergence, where the sea temperature never exceeds 4–5°C (39–41°F). They are kept cold also by the cold seas laden with pack ice and icebergs which circulate clockwise out from the Weddell Sea.

Coronation Island, which lies to the north of Signy, is the most mountainous island in the group. It is a long and narrow island, thirty miles long and five to eight miles wide. Signy Island lies five miles south of Coronation Island and is triangular, being three miles wide across its base and four miles long.

Signy is completely snow-covered during the winter months, but at the height of summer about sixty-five per cent of the land is snow-free. The snow-free areas are mainly rocky, scree moraine and poor soil, with lichens on the rocks and mossy banks around the base of the cliffs. On the island there are several freshwater lakes, which become ice-free briefly during the summer months. There are two glaciers on the island: McLeod Glacier, and Orwell Glacier.

Part Two: Signy Island

(23 November 1957 to 21 April 1959)

Life On Signy Island, Base H

Each base had to be self-supporting as it had to rely on its own resources during the long winter months. Therefore, from the last call of the relief ship in April to the beginning of the next summer season, it had to be entirely self-sufficient.

On base, alongside food, we were supplied with soap, toothbrushes and toothpaste, chocolate, writing paper and ink (which froze up during the winter), a sewing kit (which contained needles, cotton, thread and darning wool), all our clothing, cigarettes (we had over 100,000 in tins on base and only Ron smoked) and a limited supply of drink. We had one can of beer per person each week and one bottle of spirits a month between us. We saved the latter up for special occasions, such as birthdays, midwinter and Christmas.

Our base hut was called Tonsberg House. The hut was a prefabricated rectangular wooden structure measuring eighty feet by twenty-five feet and insulated with fibreglass. The main corridor ran down the length of the hut, with doors at each end. In the corridor there were two trapdoors with folding ladders to a large loft, where we stored equipment and food.

The toilet was at one end of the corridor. There were no flushing loos; instead we used old dehydrated-food tins with their tops cut open. Once they were full, if it was the summer, they would be taken out by boat and sunk; if it was the winter, they would be left on the sea ice to disappear the next spring.

On this side of the corridor, there was also the store where, amongst other things, we kept boxes of penguins' eggs; the radio room, which had the transmitters and receivers used for all our communications with the outside world, amongst the smoke haze generated by Ron's one tin of cigarettes a day; the 'met office', which had a working bench, barometer and dials from the wind recorder on the mast at the end of Berntsen Point; and the lounge, which was heated by the Esse stove fuelled by coal. In the lounge there were several chairs, a table, a record player, shelves of books, and a radio for listening to the Falklands radio and the BBC. Also there was the 'base office', which was a small office off the lounge with shelves of old reports, and all the post-office equipment. Also, off the lounge was the darkroom with all the necessary equipment for processing and printing black-and-white films. Then there was the 'survey office', which was used by two surveyors until their departure in April 1958. After their departure, I took the room over as the base office, and the old base office became a storeroom-cum-typing office for everyone to type up their various reports at the end of the season. There was also the food store, which speaks for itself! The store had shelves of tins from floor to ceiling. Then there was the equipment room, where we kept skis, tents, and other camping equipment. Finally along this side of the corridor there was the engine room, which housed two generators to supply the power for the radio and the lighting.

On the other side of the corridor there was the coal bunker (self-explanatory) and the workshop, which held a spare stove for heating the water in the bathroom next door. The workshop was well equipped with tools to make all repairs necessary. It also had a dartboard! On this side of the corridor there was also the bathroom. Inside the bathroom there was a large tank for hot water, and a bath. The outflow for the bath was a pipe which went straight out under the hut. We kept a bung in the bath so that when it was empty cold draughts did not keep blowing in. Then there was the biological lab, which had various pieces of equipment – for example, preserving jars for specimens, a microscope, etc. Jim used this room for his microphotography and to practise his clarinet. The bedroom was next on this side of the corridor. The bedroom had four sets of bunk beds and cupboards for our clothes. Finally there was the kitchen, which was divided into two sections: one for cooking and one for sitting or eating. Inside the kitchen there was an Esse stove for cooking, which we kept alight all the time. A large copper hot-water tank had been connected to the stove so that we could have a constant supply of hot water for the washing-up.

In the darkroom we all had a go at developing films and printing our own photographs. We had to log all the photographs we took with a time and date. This data was then supplied to FIDS headquarters so that if they wished they could use any photographs for their own scientific publications, for records or for calendars. Jim had photographic training and helped the least experienced with their attempts at printing.

So, with such a large hut for the six of us, and later only five, there was plenty of space for everyone to do their own thing when the conditions outside were inclement. Outside the hut we had drums of fuel for the generators, aerials for our radio equipment, an anemometer tower for the wind-recording instruments, an emergency hut with a six-month supply of food and equipment, a boat shed to store the boats in the winter, and a Met Office Stevenson screen with thermometers and recording instruments.

On Berntsen Point we had our balloon hut, where we kept cylinders of hydrogen and equipment for generating hydrogen, and beneath a trapdoor in the roof we had a pillar to hold the theodolite when following a balloon.

As the base leader, I was responsible to the Falkland Islands Dependencies Survey for the scientific programme of work to be carried out at Signy Island. Some of my responsibilities as the base leader included delegating jobs, making sure that all the work got done, encouraging others with their work and helping them as necessary. I looked after everyone's well-being and morale, and ensured that no one felt left out or overlooked. I was also responsible for all radio messages sent out and all the work reports. Being the leader also meant that I was the deputy postmaster for the base and had to reply to all mail addressed to the postmaster or base leader. Alongside this, I was also in charge of the base medical supplies, guns and ammunition.

We had more guns than we knew what to do with! There were two .303 rifles for shooting seals around the base area, two 45 revolvers for shooting seals when out sledging, one twelve-bore shotgun for the biologist to shoot specimen birds with dust shot, and one .22 rifle. We also used the twelve-bore shotgun for shooting birds, with dinner in mind. However, for this we used larger shot cartridges. With this lot and all the boxes of ammunition we could have started up our own little war if the Argentines had arrived. We had target practice once or twice when icebergs came in the bay by shooting at the pinnacles that stuck up in the air. Once we also did clay-pigeon shooting for target practice. One person would stand on top of a cliff with the twelve-bore, and another person would be on the lower ledge with a supply of wood pieces to throw for the other to try to hit.

I also had a supply of official protest forms in case any other nation tried to establish a base on Signy. Many nations lay claim to different parts of Antarctica, and this sometimes caused tensions. For example, in Hope Bay a situation arose in 1952 when the British tried to reopen a base which had been burnt down previously (more on that later), only to discover an Argentinian base had been built there. As the British went ashore, the Argentines opened fire against them. Therefore the Brits retreated to the *John Biscoe* and sent a signal to the Governor of the Falkland Islands and the Colonial Office. The Governor immediately sailed on HMS *Burghed Bay* and upon his arrival he sent the Royal Marines ashore, which caused the Argentines to hastily abandon their base. Later, sanity returned, and both bases remained at Hope Bay.

While we were on Signy, we had a board at Berntsen Point which stated that it was British Crown Land. We also had a flagpole in front of our hut, and any time a ship came to visit we put up the Union Jack. If any other nation did try to establish a base on Signy, I was to give them one of these official protest forms which stated that Signy was British Crown Land and that we would inform Her Majesty's Government of their presence. If the other nation was Argentine, due to the tensions at the time we were to be very formal. However, if it was America or another friendly nation, we were to welcome them to our base and share a few drinks, but also inform the Colonial Office that they were there.

I was also deputy postmaster (the head postmaster was on Port Stanley) on Signy Island. The post office on Signy Island was one tin box in the base leader's office containing stamps, franking equipment, a book of postal rates and a record of all letters received and replied to in the position of base leader and postmaster.

Being a stamp collector myself, I quite enjoyed this job. I replied to letters from all around the world wanting the cover of stamps, or information about us and what we were doing on Signy Island. We received letters from countries such as Australia, the USA, Poland, Japan, Czechoslovakia and England. Many letters contained international coupons for stamps and covers to be sent. One person mailed me a shoebox full of covers to be franked and sent back – it took quite a time making good postmarks on them all. Many of the people that asked us for a favour also sent us bundles of magazines, which we gratefully received. Several letters came from schools asking what it was like and questions about the geology and wildlife. However, very few realised how long it would take to receive a reply!

We were lucky as we had mail when we arrived in November, another delivery when the *John Biscoe* came in December, then again in April. We received no mail during the winter months until November, when the *Shackleton* came again. Then the next mail delivery was not until the end of summer in mid-April, when I left Signy. So, during the eighteen months that I was on Signy, we had five mail deliveries, but with gaps of five to seven months between them. Some bases, like Halley Bay and Horseshoe Island, only received mail once a year because the relief ship only visited these more southerly bases annually.

When the mail did arrive, it usually consisted of five or six sacks, and it was impossible to reply to all of the letters received as everyone was too busy and the ship was not with us for long enough. The postal accounts were checked annually by the second officer on the visiting relief ship; he collected up any monies or postal orders and coupons to return to the post office in Port Stanley, which was responsible for the Antarctic postal service. He also issued us with any new stamps to bring into use.

Here are three examples of letters I had to reply to:

1. From France, addressed to 'The Honourable the Mayor of Signy, Iles Falkland', from Union Culturelle Française, inviting the Mayor of Signy to a three-day International Friendship meeting in Paris at the Chateau de Versailles. 'Translators provided.' Unfortunately, this letter turned up three months after the meeting; otherwise it would have been a laugh to accept.

2. From Japan, addressed 'To the Honourable Postmaster, P.O. at South Orkneys'. This letter was typed on rice paper and the writer asked us about our post office and base. He asked how many people lived at Signy and if it was a seasonal base for whaling.

3. From America, addresses to 'Governor, Executive House, South Orkney Islands, British Owned Islands S.E of Cape Horn, Off Falkland Islands'. I liked this one for its elaborate address.

Midwinter party. Left to right: Brian, Gordon, Peter, Alan, George and Jim.

Signy base hut in winter.

Saul waiting for his dinner.

Weddell seal.

Hut in winter.

Coronation Island, five miles away over sea ice in Normanna Strait.

The routine reading of the temperature for the meteorology report, taken every three hours.

Water supply in winter. Perhaps this block when melted will just make a cup of tea.

Break-up of sea ice in Borge Bay – emergency hut bottom right.

Gordon Mallinson making friends with a young elephant seal.

Elephant seal.

Macaroni penguins.

Chinstrap penguin.

Adélie penguins at Gourlay.

Stanley Cathedral.

Overlooking Grytviken Whaling Station. King Edward Point top right.

I also became the base hair cutter, although this did not come under my duties either as base leader or deputy postmaster! I cut everyone else's hair, and Jim cut mine. Brian's hair was receding a bit and getting thin. He read somewhere that if you shaved all the hair off then when it regrew it would thicken up. So one day he asked me to cut all his hair off. The sight that met our eyes put us all off! We insisted that he kept his woolly hat on all the time until it regrew. Of course, when it regrew it was just the same.

We all grew beards, some more successfully than others. Alan's wasn't much more than a bit of fluff whereas George, who joined us later, grew a nice black beard. Mine was a bushy beard of brown, ginger and white hair. It got so bushy underneath my chin that I had to keep thinning it out as it got all tangled and itchy.

Gordon was the 'minister of food' and was therefore responsible for all the food supplies. All bases had enough food to last for two years in case no relief ships could reach them due to bad sea-ice conditions or other unfortunate circumstances. We kept three-quarters of our food supply (one and a half years' worth) in Tonsberg House and the other quarter (six months' worth) in the emergency hut. We changed the supply of food in the emergency hut once a year. The emergency hut was a good distance from the main hut. As well as food, it also contained other essentials, such as sleeping bags, an emergency radio and Primus stoves.

The main reason for the emergency hut was the risk of fire with wooden buildings and strong winds. At Hope Bay in the late 1940s the base hut burnt down, and the two resident members of the base hut perished in the flames. It was this base, as mentioned earlier, where the conflict arose between the Argentines and the British when the British tried to re-establish a base on the island in 1952. The only survivor of the fire was Bill Sladen, who had been camping a short distance from the base to observe a penguin rookery. When he saw dense smoke rising from the direction of the base, he returned to find the hut alight. He tried to break into the hut, but was driven back by the heat and smoke. There was no response from the two occupants to his shouts. Bill pitched his tent near the burnt-down hut at Seal Point and made daily attempts on the field radio to contact the outside world, without success. After sixteen days, Hope Bay's sledging party returned from a surveying trip to find just Bill and his tent. Later, in

1950, Bill returned south and was the base leader at Signy while he studied the penguins of the South Orkneys. Bill then lived in Canada and I met up with him in the 1990s when he came over for a Signy reunion at Whitby.

Each week we took it in turns to cook, and at the beginning of each week Gordon would issue the cook with his supplies for the week. The cook could not just help himself to any supplies; otherwise all the popular food would have got eaten first, and all the prunes and dried cabbage would have been left to last! This also made sure that we all ate a balanced diet.

Our food supplies consisted of dried and dehydrated cabbage, carrots, onions and potato. We also had powdered potato, milk and eggs; and tins of corned beef, Spam, brisket of beef and fish (kippers and herrings either in or out of tomato sauce). We had a supply of vitamin tablets to offset the lack of fresh fruit and vegetables; we also had powdered orange and lemon juice. When the relief ship visited us, it left us with a sack or two of potatoes. Occasionally, it also brought a crate of oranges or apples or the odd crate of fresh eggs from the Falklands, but none of these lasted long!

For fresh meat we shot the occasional Weddell seal for steaks, heart or liver, caught fish either by line or trap, killed the odd penguin for penguin steaks (just eating the breast), or the odd blue-eyed shag for roasting. We also collected up several hundred penguins' eggs to use in cooking. When we brought the eggs back to base, we packed them in boxes between layers of flour to keep the eggs airtight. We did eat some of the eggs, but they were not too popular as they had a rather strong flavour and the whites remained transparent when cooked. So we mainly used them for cake baking. Fish proved to be a very useful addition to the diet. We caught the fish with hook and line off the end of the jetty in the summer, and using a wire cage trap lowered though the ice in the winter, using seal meat as bait. The largest fish we caught was a six-pounder.

The cook for the week was expected to bake bread about twice a week, but in some cases the loaves were more like building bricks. Jim proved to be our best baker and at times came to Gordon's help when he had another failure! The cook was expected to plan out meals for the week using the supplies issued to him at the beginning of the week. We had a cooked breakfast, morning coffee, a cooked lunch, afternoon tea with cakes or biscuits then an evening meal. A typical day's cooking consisted of:

Breakfast – porridge, kippers, toast and marmalade.

Lunch – soup, Cornish pasties, tinned tomatoes and spud dust (made from powdered potato).

Afternoon break – jam tarts or small cakes.

Dinner – fresh fish (caught in the fish trap), peas and strips of potato (reconstituted dehydrated strips of potato) followed by chocolate pudding and sauce.

All throughout the summer, usually people were only around for breakfast and an evening meal as they were out around the island doing various jobs. They took with them bars of chocolate, a few biscuits and a flask. This made the cook's job a lot easier, except this perhaps left him to operate the radio and do all the meteorology work, which were the less-interesting jobs!

Most people had a love–hate relationship with their week on kitchen duty. They hated the beginning of their week, but loved the last day as they could now look forward to five weeks away from the kitchen until they were next on cooking duty. One of the favourite pastimes of the cook on a cold winter's day was to sit with a book in front of the stove with the bottom oven's door open (the cool oven) and put his feet in the oven to keep them warm.

Geoff took responsibility for the huskies; in December this was taken over by George when Geoff departed for his ill-fated trip to Horseshoe Island. The dog man was responsible for making sure that the dogs were in good health by feeding them, which entailed shooting the occasional seal and cutting it up into manageable pieces and making sure the dogs were well exercised. He was also responsible for repairing their harnesses when necessary. We only had six huskies, which we kept on a long wire span pegged out over the rough rocky ground. We kept each dog on a lead attached to the span. The length of the leads stopped the dogs from reaching the next one, to prevent any fights breaking out. However, when a lead broke, fights did sometimes break out. Then it was all hands on deck to separate them before too many injuries were inflicted while watching out for our own hands and legs. When loose off the span, except when under strict control pulling a sledge, they had a bad habit of wanting to pull one another limb from limb.

Alan was our local bird man (ornithologist), and it was his job, with the help of others, to ring the entire giant-petrel fledgling population on the island. He also had to ring many other birds, including skuas, penguins, prions, Wilson's petrels and black-bellied storm petrels. This meant that he was in charge of keeping the ring register up to date. Alongside this, Alan made regular bird counts to monitor the island's bird population. I took over from Alan a year later when Alan went on his ill-fated trip to Admiralty Bay.

Alan was also a keen amateur radio operator (ham), and his call sign was VP8DT. In the winter he regularly used to try his hand at 'hamming' and contacting people from all over the world. If the weather conditions were good, then we could easily hear from the USA, South Africa and Great Britain. On these occasions, he used to come up with his call sign and then sit back and listen to all the other hams trying to contact him. A QSL card from the Antarctic was a prized possession amongst operators as there were so few hams in the Antarctic. A QSL card is a card that the operators exchange with each other once they have made contact, which states the time of contact and the strength of the signal. Some Americans must have had very powerful transmitters as they used to drown out all the other operators with their signals. Alan used to wait and try and contact the weaker signals as they obviously had a harder task trying to make themselves heard above the more powerful transmitters.

Jim was my deputy and our biologist. He collected specimens to send away and using a microscope he photographed the microorganisms in water he had collected from the freshwater lakes near Pipeline Beach. He was also our 'base iceman', which meant that he made observations on the sea-ice situation around the island. He also observed the amount and form of the pack ice in the summer and the extent of the sea ice in the winter. He had to complete ice observation, weather permitting, weekly from the highest point (Tioga Hill) on the island. When ships were near, he had to complete ice observation daily. When the sea had frozen over in the winter, he had to measure the thickness of the ice by drilling holes in it and measure the sea temperature beneath the ice. He was helped in these duties by all the other staff.

Brian, who was a very adaptable chap, was our general handyman-cum-general-assistant, who helped everyone else out with their work. He was good at shooting and undertook most of the shooting tasks, from killing seals to feed the dogs (or us) to killing birds for the pot. He also shot specimens to be skinned or preserved in formalin for the official biologist, Fergus O'Gorman, my cabin companion on the journey down, who had spent his first eighteen months elsewhere and arrived on Signy the day I left in 1959.

Gash duty was a daily job that was taken on in turns by everyone, except the cook for that week. We had to sweep out the living room and tidy it up, clean out and stock the fires in both the living room and the workshop, and make sure that the coal scuttles were refilled. It also meant sweeping down the central corridor and clearing the snow away from each door as otherwise the doors often got drifted up and pack snow got in the cracks in the frame, which preventing the door from shutting. Also, if the sack of empty tins was full, we had to dump them. In the summer we dumped them down in Gash Cove, or out in the bay from a boat. In the winter we dumped them in a heap on the sea ice way out by Billie Rocks, to disappear with the sea ice when it broke up in the summer.

Another task was filling up the water tanks with snow, ice or water. In the summer, when the temperature was above freezing, which happened at times for about three months of the year, a stream of water from melting snow was piped into the four drums by the back door, where it could easily be brought into the hut in buckets. The rest of the year, we either cut snow blocks or used ice from the bergy bits off the glacier. The snow blocks could be cut quite easily from the drifts around the hut later in the winter when there had been a considerable snowfall. But before the drifts had built up, ice had to be broken off from bergy bits in the bay or from the face of the glacier in the shallows. Both ways a boat had to be used.

The day before we were on gash duty we could have a bath and do as much or as little washing as we liked, but the next day we would have to fill the water tanks up to the top. When the water or snow was in short supply, we went sparingly with the water, but in the summer, if the melt stream was running, we made the most of it. When cutting snow blocks, we cut the largest blocks that we could carry. Then we would stagger in with them and lift them into the water tanks, only to see the water level rise a small fraction! Other times, when carefully carrying a large block into the hut, it would suddenly disintegrate in our arms just as we were about to lift it into the tank and we would have snow all up our sleeves and over the floor. Chopping pieces off bergy bits was a two-person job – one person kept the boat steady while the other chopped away with the axe. We would hope that a huge piece of ice would not break off and swamp the boat or turn it over!

The last duty of the skivvy (the person on gash duty) was to help the cook and to do either the washing-up or the drying and to empty the gash buckets at the end of the day.

On Saturday mornings everyone lent a hand to scrub the floors of the bathroom, living room, toilet, kitchen, wireless cabin and met office. In the summer, we would scrub the bedroom too. We would not scrub the bedroom in the winter because the temperatures were below freezing nearly all the time. When this happened, the floor got covered in a sheet of ice, and the moisture from the floor would get into the mattresses, causing them to freeze to the beds. Ice formed in other rooms just the same; but in all but the toilet, as there was some heating or warmth from other rooms, the floors dried out during the day. We noticed that ice formed on the floors when the air temperature outside was -14ºC (6.8ºF) or below; and as the average temperature throughout the winter was below this figure, ice on the floor in the winter was very common.

During our time on base, we were allowed one air letter in and out each month. We sent it by radio from the base to the Falklands headquarters, where it was typed up by the office staff and sent by sea mail to Montevideo and then by air to England. Letters from home came the reverse way back. In each case, the maximum number of words was 100. We could never write or receive personal letters as our radio operators read them all, as did the radio operator in the Falklands and the typist in the office.

Once a year, each member in the Antarctic received a family message from home in a programme broadcast to the Antarctic by the BBC called *Calling the Antarctic*. The half-hour programme was beamed to the bases to keep us in touch with family messages and news of interest.

In my eighteen months at Signy there were only two months when the average temperature was above freezing. The first was in

January 1958, when the average temperature was 0.3°C (32.4°F); and the second was in January 1959, when the average was 0.2°C (32.4°F). The annual mean temperature for 1958 was -4.9°C (23.2°F). The lowest mean monthly temperature was May 1958, when it was -14.3°C (6.3°F). The lowest temperature we recorded was -32°C (-27°F), in July 1958. On the other hand, we once saw the temperature rise in January 1953 to 11.7°C (53°F). When a cold-front weather system passed through us in winter, the temperature often dropped by fifteen to twenty degrees Fahrenheit in a matter of two to three hours.

Along with the low temperatures it was also very windy, with an annual mean windspeed of over fifteen knots. Our windiest month was September 1958, when the mean wind speed for the month was twenty-four knots. That month also had twenty-two days on which we recorded gale-force winds. In a year, we recorded 113 days when we experienced gale-force winds. The highest gust we recorded was 115 knots on 9 August 1958, and in five different months we had gusts of over eighty knots.

So, putting these two factors of wind and temperature together, we had to be very careful about dressing adequately. The wind could spring up very quickly in the winter with low temperatures. When out, we were warned to keep an eye on each other for signs of frostbite – for example, a white patch appearing on the cheeks or tip of the nose. If we did see signs of frostbite, we were to rub the affected area with our hands to warm the area and stimulate circulation. Fortunately, none of us suffered from frostbite.

We could easily notice the effects of the wind due to the snowdrift that built up outside our hut. However, the hut had been raised above the ground on small brick pillars. Therefore, the wind could blow beneath the hut and form a wind scoop. This meant that we could walk between the wall of our hut and the drift. This made it ideal for cutting snow blocks out of the drift for our water supply. The wind also blew the hard snow into ridges and scoops, called sastrugi, which were difficult to travel over with a sledge. The wind also blasted the sea-ice surface, polishing it and making it very slippery and smooth, which meant that it was very difficult to walk over in any wind. Usually the sea-ice surface had a layer of snow, and travelling over it was like crossing a frozen field covered in snow.

When on meteorology duties, we did twenty-four hours at a time, from 6 a.m. through to 6 a.m. the following morning. After making the first observation and sending it by radio to the radio station in the Falklands, we then stoked up the fires and got the kitchen stove hot for when the cook got up. In the winter, we had to continually clear snow from the instrument screen; otherwise the equipment would get clogged up with snow and stop working. Also, we had to de-ice the sunshine sphere of rime so that we could record any sunshine. We also had to send balloons each day when there was no low cloud to obtain wind speeds and directions at various levels. We filled the balloons up with hydrogen so that the balloon would rise at either 500 or 700 feet (152–213 m) a minute. We had two cylinders of hydrogen, and when all that was used up we generated hydrogen with a low-pressure generator, which in the winter often took about half an hour to fill the balloon. We followed many balloons for over an hour, when they would be at 30,000 feet (9,144 m). We took readings every minute, recording the elevation and azimuth of the balloon, then when we lost the balloon, or it burst (or we got too cold!), we had to sit down indoors with a slide rule and work out all the wind speeds and directions for every 1,000 feet (305 m). To protect us from the weather when following the balloon, we had a balloon hut made from an old packing case with a trapdoor in the top and a pillar in the middle to set the theodolite on, so we could stand in the hut with just our heads sticking out.

On Signy Island, we had many interesting clouds formed by the passage of air over the mountains on Coronation Island. I took bearings on the clouds when they formed, and sent up balloons to find out the wind conditions when the clouds formed. I wrote a paper about my findings on the formation of these clouds in the *Meteorological Magazine*.

In the winter, we sometimes saw mirages, usually of icebergs on the horizon or of islands upside down. Also, we saw halos and mock suns; and one day, when there were clear blue skies, the air was filled with ice needles suspended in it. You could only see them if you were looking into the sun. Then they sparkled and twinkled in the sunlight to produce halos.

Summer On Signy Island, 1957–58

While we were settling down on Signy, drama was not far away. After disembarking the Powell Island party, the *Shackleton*, while making its way north of the South Orkneys in heavy pack ice, was holed on the port side below the waterline in a forward hold.

All the cargo was moved to the starboard side of the boat to try to lift the port side up a bit so that work could be carried out to shore up the hole. The ship's crane was also swung out over the side with the unloading barge on it to help.

The second officer, Tom Woodfield, and Dennis Wildrich worked for a long time in freezing-cold water to patch up the hole from inside. The Doctor sent several of the helpers up out of the hold after they had been down there for about an hour, so that they would not suffer too much. People were continuously making hot cocoa in the FID'ary. All the ship's lifeboats and its motor boat were loaded up with the ship's papers and several people. All the people working on board the ship wore life jackets. The first officer, Tom Flack, did all the organising above and below decks. The bosun was drunk as usual and so therefore of no use! A lot of the ship's cargo had to be ditched to help movement below deck.

The captain, Norman Brown, decided that if the ship was not on an even keel by 4 p.m. that afternoon, then they would abandon it as there was a chance that the ship would roll over. Varying reports suggested that it would have taken six inches to two feet (15–60 cm) of water to roll her over. Dramatic as it may sound, the Captain was in the wardroom with the chief engineer, Jack Richardson, at 3.50 p.m. when he noticed that the curtains were hanging almost straight again. In other words, the ship had righted herself.

After shoring up the hole, cement was poured in behind the woodwork to try and block the hole, but work had to be done over the side under the waterline. At first, nobody could be found to do it, so David (second steward) volunteered to go over and do the patching-up.

One of the whaling boats from Leith was the first to arrive on the scene of the accident in answer to the SOS call put out. The whaling boat stayed with the *Shackleton* until HMS *Protector* arrived and accompanied her to Stromness in South Georgia, where she was dry-docked and repaired.

As soon as they got to South Georgia, the Captain went on board the *Protector* and got drunk and remained there for about twenty-four hours. He resigned his job on returning to Port Stanley, saying that he did not think the ship was suitable for the job she had to do. It was reckoned that if we had not unloaded the two buildings in South Georgia and the stores at Signy before the accident, she would have sunk because there would not have been room in the hold for the cargo to be moved around to enable the men to patch the hole up.

Back home, it must have been a worrying time for my parents as it was not known at the time that I, along with my colleagues, had been put ashore at Signy. My parents would have seen 'S.O.S DRAMA AS BRITISH SURVEY SHIP HITS ICEBERG', 'ANTARCTIC RESCUE RACE' in the *Daily Express*, and 'ANTARCTIC S.O.S – SHIP SINKING' in the *Evening Standard*.

My parents received a telegram from the FIDS headquarters, Crown Agents, in London. Oh, for modern-day communication in 1957! My parents and all the other parents would have known where and what was happening much more quickly. My parents would have known that I was safely ashore on Signy.

As for us on Signy, only a short distance away, the first we heard about the accident was via the BBC Overseas Service two days later. Radio conditions had been very poor, and we had not been in contact with the headquarters in Port Stanley for several days. Also, Gordon still had not got to grips with the radio gear on the island. We had had no contact with other bases or the *Shackleton* during all this time. Today you can pick up a phone in the Antarctic and phone home via a satellite – instant contact with the outside world! However, I am glad that we had not got these facilities as it would have made us seem a lot less remote and more accessible – not a true explorer's feeling! If we had a problem, we had to sort it out ourselves; today it's all too easy to phone home!

The *John Biscoe* visited us in December, bringing George White (diesel mechanic) to take over from Geoff Stride, as well as Cecil Scotland and Derick Skilling, who had been on Powell Island for a month. The latter two were just staying until the end of the summer to help us out with the bird ringing as they had already completed two years on the bases.

All throughout the summer almost every day parties left the base, either by boat or on foot, to survey, ring birds, count seals, make ice observations or just explore. Some days just one person would be left on base and he would act as a wireless operator, meteorologist and cook until the others returned.

We made regular seal counts every time we visited beaches where seals came to breed or moult. Well over 4,000 elephant seals were residents of Signy Island in the summer. There were only a few dozen Weddell, crab-eater and leopard seals. Fur seals were only occasional visitors to the island; they had almost become extinct due to the fur trade in the past. Whenever we spotted any fur seals, we would make a note and keep watch on them. Now, I believe that they have multiplied to quite large numbers.

In the summer, bird work took many forms. We had to ring all the giant-petrel fledglings in March and April, which amounted to about 1,500–1,800 birds. As our first summer on Signy was part of the International Geophysical Year (IGY), we had to put a special blue-and-white plastic band on the same leg as the ordinary aluminium bird ring and a coloured spiral ring on the other leg. The coloured spiral ring indicated to future bird men the age of the bird as a different-coloured spiral was used each year. The blue-and-white plastic band indicated that the bird was from the South Orkneys. This was a tiring and dirty job. The birds have a wingspan of almost six feet (180 cm) which means they are awkward to handle. Also, when we caught them, they regurgitated all the food that they had been fed recently, usually down our anorak and trousers – a very smelly oily concoction of mainly krill and squid. We had special waterproof gear solely for this job! We did a lot of the ringing while camping on the west coast of the island, but some days up to four of us boated or walked over via the ice cap when the weather was too rough for boating, and we would spend the entire day ringing birds. This work brought some spectacular recoveries: within a month of leaving the island, some birds were recovered in South Africa and a few were recovered less than three weeks after leaving Signy in the Fremantle and Perth area of Western Australia, 8,500 miles away! Adult birds did not seem to undertake this great migration, but remained nearer their breeding grounds instead. The young birds, when they are old enough to migrate, leave and do not return until they are four years old to breed in the same area as where they were born. How do they manage to find their way back? Some returning adults were ringed ten years before, when the base was first set up.

We also ringed penguins on their flippers and around their legs. On the island, there were three main types of breeding penguins: Adélie, chinstrap, and gentoo. There were also the odd one or two pairs of macaroni penguins that bred. Occasionally the odd king penguin visited, but none nested in the South Orkneys. It was impossible to see the leg ring because of their feathers, which meant the flipper ring proved very useful in making recoveries. With good eyesight or a pair of binoculars, the ring could be read without having to catch the bird first. The penguins returned to the same rookery each year with the same partners. We also web-marked whole rookeries of Adélie, chinstrap and gentoo chicks in the North Point area as a quicker method of marking all the birds from the same area. We ringed many Cape pigeons, snow petrels, Wilson petrels, and prions at their nesting sites near the base to establish breeding patterns and lifespans.

An earlier bird man (Lance Tickell) was writing a paper for his PhD on the life cycle of the prion, and he asked us to locate a number of nests. Then, when the eggs hatched, we were to weigh the chicks every morning and evening until they had left their nest about six weeks later. Also, we were asked to put some fine pieces of wood up in front of their nesting burrows so that we could see if the parents had been back to feed their young. This was a tedious job as we had to climb up and down the cliff near the base in all weathers with a pair of fine scales in a box. Then we had to put our arms way down a nesting hole to extract a vomiting chick to note its weight and see whether the parents had been back to feed it. On several occasions, the nest got snowed in, and we had to dig them out by hand first.

Another task we did was mapping and putting concrete markers by all the sheathbill nests around the penguin rookeries at Gourlay. Sheathbills are scavengers that eat eggs or dead birds as well as picking over seal carcasses.

During the IGY there was a special study of the skua, which meant that we had to map all their nests, of which there were about sixty. We also had to ring as many as possible of the adults with a combination of coloured rings so that we could tell which nests they belonged to and could note where they fed compared to where they nested. We had to put an aluminium ring plus a plastic blue-and-white band on one leg to indicate that the bird was ringed on the South Orkneys, and on the other leg we put a combination of different-coloured spiral rings to indicate which nest they came from. Several birds had up to three spiral rings on their legs. Any time someone saw a skua with coloured spiral rings on it, he had to make a note of where and when he had seen it and enter it on to a wall chart which detailed the birds' movements around the island. We also had to ring as many other skuas, non-breeding and youngsters, as possible. To help with this we constructed a trap in front of the hut baited with a lump of seal meat. The door of the trap was operated by a piece of string from the hut so that we could look out of the window and if we spotted a skua in the cage we would pull the string to shut the cage door. Then we would don a pair of gloves and go out and catch the bird and ring it with an aluminium and blue-and-white plastic band. To catch the skuas that were nesting we used a net stretched between two poles as when anyone got near to their nests the skuas would attack him, swooping down and clipping the top of his head with their feet. As the skuas dived, we would duck down and swing the net above our head and end up with one squawking bundle of a skua in the net to be ringed.

Sporadically during the summer the odd whale was seen swimming in the Normanna Strait between Signy and Coronation Island. However, we had no close encounters with whales while out boating. The only sea animals we encountered while out boating were seals, of which the leopard seal was the most inquisitive – they kept coming back to look at us! I expect most other sea life was disturbed by the sound of the outboard engine.

During January, Doug had been camping at North Point for about a week with Brian to complete the final triangular survey of

Signy, linking the island with Coronation Island to the north. On 4 January I was due to join Doug. As there was so much pack ice in the bays, I could not get by boat all the way round to North Point, so Rob took me to Pipeline Beach, and I then walked to North Point via the col between Jane and Robin Peaks. Earlier in the day, Alan and Cecil had managed to get through to North Point by boat to do some bird ringing. However, due to the ice moving in, which curtailed my journey, they had to leave their boat at North Point and walk back accompanied by Brian via the ice cap and Garnet Hill, where Alan half disappeared down a crevasse. After dropping me at Pipeline Beach, Rob got stuck in the ice with his boat. Upon seeing his plight, Derick and George went to his rescue with the third boat, only for them also to get stuck! They all finally managed to floe-hop, pulling their boats over the floes to reach Shallow Bay, where the boats had to be left to be collected when the ice conditions improved. They then all walked back to base via the Stone Chute.

The tent that Doug and I stayed in was a double-lined two-man pyramid tent, which meant that it was like a tent within a tent. In the tent, we had two sleeping bags on sheepskins, and between the two sleeping bags were two sledge ration boxes. One of these boxes contained our food supplies (all dried and dehydrated) and the other contained essentials. On top of this box we balanced a Primus stove to do our cooking. One luxury we had with us was a radio to listen to Gordon at the base or the BBC Overseas Service when the reception was good enough.

The tent was quite near an Adélie penguin rookery; and with it being daylight most of the day and night, their chattering kept us awake at times. On one occasion we kept hearing a thump on the guy ropes on one side of the tent, and on investigating we realised it was caused by a stupid penguin which kept walking into the guy rope, falling over and then doing it again! It looked so silly – he did not seem to be able to work out that he only had to walk around it!

The first day at North Point started off fine and calm, so we climbed to the top of Robin Peak to take some triangulation readings. I felt like a packhorse carrying a heavy theodolite and tripod up to the top of the peak. Later I sat on the theodolite box while Doug called out readings to me. The weather clouded up later in the day, and it was snowing quite heavily by the evening. So we spent the afternoon mapping the penguin rookeries and counting the nests.

The next three days were dull, with cloud sitting on the hilltops, strong winds and snow at times. This weather was not good for surveying, but Doug managed to finish mapping the Penguin rookeries, and I counted their nests. I also brushed up on my trigonometry by helping Doug with his calculations.

Around North Point and down the west coast of the island there were many giant-petrel nests, and at this time of year their eggs were starting to hatch. Therefore in a few months' time I would be back to help ring all the young birds. All the time we were camping we kept hearing rumblings from glaciers on Coronation Island and icebergs breaking up. The sea was full of icebergs, between 400 and 450 could be seen from the top of Robin Peak.

Finally, on the fourth day the weather was kind to us, and we went to the Spindrift Rocks area to finish off the survey work and map and count the penguins there. Later that day, Derick and Brian turned up by boat. The ice by now had broken up more and moved out, so there was an opening to boat through. That evening we packed up and made an uneventful return to base.

Doug was the most experienced camper, having spent well over 400 days of his two and half years in the field living in a tent surveying the South Orkney islands. It was a great experience to camp with him, and I was able to learn all the tricks of the trade, from cooking without getting out of bed to keeping your clothes dry, doing the more basic functions without getting wet or your trousers full of snow. During the summer, as well as myself and Brian, Alan and Jim also managed to get away from base and spend some time camping.

Over the summer, George thought it was a good idea to move the old oblong gash barge, which the whalers had used to dump whale waste out to sea, from the site of the old whaling station. The reason George thought this was a good idea was because it would give us a lot more space on the whale plan in front of the hut to pull the boats up. Doug bet George that we would not be able to do it. So in all his spare time over the first summer, George, with various help from others, cleared up all the old scrap rubbish from around and in the barge. With a sledgehammer, he drove the old whaling harpoon head under the barge to lift it. He also made several boating trips to Pipeline Beach and brought back pipes to slide under the barge to act as rollers. Then, one day in March, we slid the pipes under the barge and laid out other pipes in front of the barge, and with a push and a tug the barge was off, rolling down into the sea.

Once afloat, the question was what to do with this barge, as next to this barge there was a sunken barge which earlier members had built the jetty to. We had to decide whether to take this barge out to the middle of the bay and sink it, or sink it next to the sunken barge at the end of the jetty and fill it with rocks to make the jetty longer. We decided on the latter, so the barge was temporarily moored to the jetty while we opened the inspection covers on the buoyancy tanks. It took a lot of hard work and penetrating oil to move the old rusty nuts holding the traps of the tanks – especially as the barge had been lying idle and rusting for forty years or more. Finally, we prised them open, flooded the tanks and filled them with rocks to stop the barge from moving before we put in more rocks around the edge to prevent it from slipping. We laid timbers down from the existing jetty out on to the newly sunken barge, and when we completed it we realised that we had almost doubled the length of the jetty. George won his bet!

An extra job we had in the summer months was making tide readings every three hours using a measuring pole attached to the sunken barge at the end of the jetty. The readings were to work out the mean sea level for the island. We also helped the surveyors on clear nights to take readings on the stars to get an astro-fix to pinpoint Signy on the map. Other jobs we did during the summer were to completely tar the roof, creosote the walls and paint the doors and windows.

Winter On Signy Island, 1958

When Cecil Scotland, Derick Skilling, Doug Bridger and Rob Sherman departed on the *John Biscoe* on 15 April, it was farewell to all visitors and shipping for the next seven months. The remaining six of us were left on our own to face the winter and to cope with any problems that might come our way without the experience of others who had previously spent a winter in the Antarctic.

Within a week of the *John Biscoe*'s departure, the sea started to freeze in the bay in front of the hut to a thickness of three inches. By 2 May the sea had frozen to the horizon, and we could not see any open water. Occasionally in stormy weather the sea ice broke at a distance or the odd iceberg could be seen being carried along by the wind and tides. In these early days of the sea-ice formation, we did not venture out of the bay on the ice as it was still liable to break up and crack with the tides and currents underneath.

On 7 May, George, Alan and I went overland to Gourlay on skis to see if there were any signs of life around the peninsula, but we could not see any penguins or seals. Gordon, after his radio sked (a radio schedule), thought he would catch us up, so he clipped on his skis and went out over the newly formed sea ice to Gourlay. We were amazed when he arrived, as there were cracks around the headlands in the ice and the ice was not very thick. He said that there were one or two gaps and the ice did go up and down a bit as he went over it. He was a bit worried, but he thought it would be fine as he assumed that we had gone that way. How we never lost our wireless operator I do not know! Up to that day, nobody had ventured outside the bay on the ice.

During the early part of May, we saw long processions of elephant seals leaving the island from North Point and the Elephant Flats area, travelling over the sea ice to look for open water. I don't know how far they had to go, but it must have been a long way.

66

In the winter bird life was sparse. A few sheathbills scavenged around the base and we saw the odd snow petrel and a few blue-eyed shags. There must have been some open water somewhere for them to feed, but we could not see it. The birds returned to Signy with the break-up of the ice, at the start of the breeding season, when we had about fifteen different species nesting.

By June the sea ice was established and in very good condition for travelling over. It was fortunate that there was no pack ice about when the sea ice formed. If pack ice had been present, it would have made travelling over it difficult because the ice would have been very bumpy due to the packs that were frozen into it.

Our small team of dogs – Garth, Sampson, Saul and Dyke – were taken out on to the ice and trained to pull a sledge. George and Brian spent many hours training and encouraging them and breaking up the odd fights that they got into. We then started to make journeys with the dogs over the sea ice around the island, and everyone went at least twice across the five-mile frozen Normanna Strait to Coronation Island. The trips were simply for our enjoyment, to exercise the dogs and to take photographs.

We kept the dogs on a long wire span, and throughout the winter this had to be raised several times as the snow depth increased. The dogs were fed every other day on a large lump of seal meat and occasionally a bar of pemmican or a penguin. Before the winter set in, we shot several seals to stockpile as food for the dogs during the winter. With the very low temperatures, the dog's meat had to be chopped up by axe and then given to them as a solid lump. They found their own ways of defrosting their meat, either by lying on it or peeing on it! Once, Garth and Saul had a fight which left Garth with a badly cut foot. We brought him into the workshop, where Brian cleaned the cut and stitched it. Garth was kept in the boat shed until the wound healed so that he would not get dirt or ice in the cut.

Once or twice during June, we saw the odd seal come up through a crack in the ice. If it was a Weddell seal, we would grab a gun and shoot it for fresh meat for the dogs and ourselves. On one of these occasions, George, Jim and I collected a seal from Gourlay on a sledge pulled by Garth, and when we crossed a tide crack coming down from Polynesia Point my skis sank into it and almost disappeared, but I managed to scramble out. The journey was hard going as we had recently had a heavy snowfall with strong winds, and in places the soft snow was over two feet (60 cm) deep in drifts on the sea ice.

All through June and the beginning of July we cut holes in the sea ice in the bay and put our fish trap down. We had many catches of ten-to-twenty fish weighing between one and four pounds. However, later in July the ice became too thick to chop the hole through to sink the trap, although Jim carried on boring holes in the ice to measure the thickness and take the sea temperature under the ice.

On 21 June to celebrate Midwinter's Day we cooked a special meal and opened various goodies, including cheese straws and a special tin of chocolate biscuits. We also brought out some extra drink that we had been saving up for the occasion. We all dressed up for dinner and Jim photographed us all at the dinner table. We spent the evening playing games and records and trying to answer a quiz that Jim had prepared. Brian had a bit too much to drink and went for a walk around the hut in his short sleeves with temperatures well below zero. I remember him lying flat out in a snowdrift laughing his head off – we had to carry him inside!

Unfortunately the month of June also brought us bad news: on 3 June we received a message from the FIDS secretary asking all bases to listen out for a sledging party from Base Y which had been missing since the sea ice broke up further south.

We had a radio sked with Base Y on 13 June and found out that the party was still missing. We also heard that Geoff Stride, who was with us on the *Shackleton* travelling south and was our diesel mechanic at Signy for a month until George White arrived on the *John Biscoe*, was one of the missing sledge party, along with Dave Statham and Stan Black, both former Signy Islanders. We had taken over from Dave and Stan on our arrival – they had gone on to Powell Island for a month before being picked up and taken to Horseshoe Island on the *John Biscoe*. The sledge party had been missing since 27 May. Both Base W and Base E had sent out search parties.

We heard on 24 June that two of the dog team belonging to the missing party had been found in good condition by one of the search parties.

Then by 28 June they had found seven of the dog team, and by the beginning of July ten of the dogs. However, some of the dogs did not have harnesses, and four more of the dogs were still missing.

The search concentrated around Dion Island, Faure Island and Cape Alexandra, north and south of Base Y.

Then on 18 July we received this message from the Governor of the Falkland Islands to all bases:

> It is with great regret that I have to inform you of the loss of a sledge party consisting of Black, Statham and Stride. They left base on 27th May for the Dions with intention of camping first night on Pourquoi Pas Island. Gale arose around midnight on the 27th May and when it subsided base 'Y' saw ice had gone out. Search made of Pourquoi Pas as soon as it was safe but nil found and since further searches have been made by parties from 'Y', 'E' and 'W' in all possible areas without success. All but four of their dogs have turned up. It must now be assumed that they unhappily decided to camp on sea-ice first night out and were caught in break-up.

We later heard that the Argentine base found some more of the dogs.

In the middle of winter, when we could not get out much, Brian, not realising it, upset Alan and George with his singing as he walked up and down the corridor (he couldn't sing very well!). We were so on top of each other in the hut, even though it was a large hut with many rooms, but no one could escape his voice as he went backwards and forwards up and down the corridor; so I had to have a diplomatic word in his ear, to suppress his voice a bit. He was most peeved, but after thinking about it agreed to try and curb himself.

Brian was also a bit of a fresh-air fiend, although he liked his warmth as well. I remember him once opening the bedroom window at night and then piling more clothes on his bed to keep warm. In the morning he woke up and found it had been snowing hard and his bed was covered in snow.

Only one of us, Gordon, had brought a hot-water bottle, which made the rest of us rather envious when the temperatures dropped in April. After leaving his bottle in bed all day, by the next evening, when it came to be filled again, he often found it frozen solid, so he used to thaw it by placing it on the rack over the kitchen stove.

Ingenious George, not to be outdone, made his own bed-warmer from a biscuit tin, which he wired up to the mains supply. In the tin he put a low-wattage bulb. When the generators came on in the evening to charge the batteries, he wrapped his pyjamas around the tin and placed it in the middle of his bed; so not only did he have a warm patch in his bed, but he also had some nice warm pyjamas. Brian and I both copied this. The only thing was that Brian fancied a really warm bed, so unbeknown to anyone else he put a 150-watt bulb in his biscuit-tin bed-warmer. That evening Brian switched his on for the first time, and a little while later I switched mine on too. I found that there was a smell of burning in the bedroom – it was Brian's bed! Quickly his bed-warmer was switched off, and the smouldering bed stamped out, but not before Brian's pyjamas were altogether destroyed. There was also a hole in his sheet, and three blankets were scorched. Now, there's how to warm your bed or burn your hut down in one easy lesson!

After this incident, which we omitted from the base diary, but we did not from our personal diaries, that type of bed-warmer was banned. In the end, most of us ended up putting firebricks in the oven to get hot in the evening, then wrapping them in newspaper and stuffing them into a boot sock to put into our beds. This proved to be a very successful and safe bed-warmer. It was often still warm the next morning.

On a still night in the winter, we could hear the occasional crack and bang from the sea ice around the shore as it rose or fell with the tide. We used to say it was the dead whalers knocking to come in.

In the middle of winter I remember once accompanying Jim to Tioga Hill to make an ice observation when it was very cold. We took a bar of Cadbury's toffee chocolate and a can of beer each inside our coat pockets. On reaching the cairn on top of Tioga Hill, we sat down on the snow to have our snack only to find the beer had frozen solid and our toffee chocolate bar was as hard as concrete!

Also during the winter the Stone Chute near the base had a permanent snowfield at the side, so we used this as our Cresta Run. We took a small sledge to the top, then we would come hurtling down and out over the sea ice. A ridge developed around the tide crack, which was caused by the ice rising and falling with the tide. When we went over the tide crack, we had to hold tight to the sledge. Gordon did not realise this the first time, and when he went over the crack the sledge came up and hit him in the face, which caused two nasty cuts.

When it was snowing, sometimes drifts formed against the door, and on opening it to go out and make an observation we were met by a solid wall of snow. Then we would either shut the door very carefully so that the snow did not come tumbling into the hut, and go out by the door at the end instead, or we would get a shovel and poke it through the wall of snow and dig the doorway out. We would be very careful not to leave any packed snow around the door frame; otherwise the door would not shut properly. I can remember tumbling out of bed one morning to do the 6-a.m. observation, pulling my clothes on over my pyjamas, staggering out through the door and going flying head first into a snowdrift. I had tripped over a sleepy elephant seal which just so happened to be lying in the doorway, I don't know who was most surprised, the seal or me! It certainly woke me up!

During the bad weather in the winter we busied ourselves in various ways. We redecorated various rooms, lay new lino in the kitchen and bathroom and made repairs to the sledges and equipment. We also made slides of snowflakes, which were sent off to the scientific bureau. For leisure, we spent a lot of time playing darts and cards or in the darkroom developing and printing films. However, whenever the weather eased we got outside to do various tasks, such as releasing weather balloons, attending to the dogs, making ice observations and conducting seal and bird counts.

Several times we had trouble with our wind-recording equipment icing up, and Alan and I spent many hours in freezing conditions up at the top of the anemometer tower de-icing, rewiring and renewing the directional transmitter. Thirty-three feet above the ground was not my favourite place to be with frozen fingers.

The FIDS also wanted building research work done during the winter, to see what conditions were like in the huts and to see if insulation could be improved in the future. The work consisted of taking temperatures at floor, head and ceiling height in various rooms in the hut, including the loft, and relating these figures to the conditions outside. I also had to note the height of snowdrifts around the hut and the accumulation of snow on the roof. This took two to three hours to do each day.

By early August, although there was no sign of the ice breaking up, 200 to 300 giant petrels were seen on the west coast and over 500 snow petrels. On 12 August we could see open water off Gourlay; but as ice conditions between Signy and Coronation Island were still good, we continued our trips across the Normanna Strait. About this time, we spotted the first Weddell seals on the west coast of the island, where they return to have their pups each year on the ice. However, there was no open water on the western side, so the seals had to come up through cracks around grounded icebergs.

Here is an extract from my diary from the week 4–10 August 1958.

4 August

Windy day, wind west-north-west with a daily mean speed of 25 knots. Gale-force winds from 0300–1100. Highest gust 64 knots. Temperature started off 3ºC (37.4ºF) and dropped to -7ºC (19.4ºF). Lowest pressure 959 mbs. The temperature drop meant that surfaces everywhere were very slippery. I sorted out stores to be exchanged with the emergency hut when the weather improves. I took temperatures all around the hut at floor, head and ceiling levels for Building Research Programme. I spent time in the darkroom in the evening printing some negatives. Listened to *Goon Show* on BBC Overseas Service.

5 August

Wind north-west with a daily mean speed of 37 knots. Gale all day till late evening – highest gust 81 knots. Temperature 0–12ºC (32–54ºF). Snow and drifting snow all day. The visibility was very poor. Lowest pressure 967 mbs. Even with the gales and snow in the afternoon, all but Gordon went out over the sea ice to Billie Rocks, where we shot and gutted a Weddell seal to feed the dogs. It was the first seal we had seen in weeks. Took usual temperatures around the hut, listened to *Calling the Antarctic* from the BBC in the evening. They played a record for me by Louis Armstrong – 'Ain't Misbehaving'. I then heard that they will be broadcasting my message from home next week. I listened to Base F playing Benny Goodman records late in the evening. Did early half of met duty today and Brian did the late shift.

6 August

Wind light or calm all day. Temperature -16ºC (3.2ºF) to -21ºC (-5.8ºF). Lowest pressures 983 mbs. I followed a balloon to 8,000 feet and worked out winds every thousand feet in the afternoon. The weather was fine and sunny except for a few flakes of snow around midday. I continued taking temperatures around the hut and did late shift on met. George, Brian and I with the aid of dogs and sledge took new supplies via the sea ice to the emergency hut and brought the old supplies back. Then again with the dogs and sledge I collected the carcass of the Weddell seal shot yesterday out by Billie Rocks.

7 August

Wind north-west at a mean speed of 32 knots. Gale or near gale all day. Highest gust was 58 knots. The temperature began at -17°C (1.4°F) and gradually rose all day to -1°C (30.3°F). Snow and drifting snow all day. Lowest pressure 981 mbs. I took the usual temperature measurements in the hut. In the afternoon, Jim and I walked to the west coast of the island for a sea-ice observation and a general look around for seals and birds. Still solid ice for as far as we could see, but we saw about 200–300 giant petrels and 500 snow petrels. They must be coming back early for summer. Also, I saw twenty blue-eyed shags at their nesting site at North Point. There was one Weddell seal on the ice at North Point. Brian and I stacked more coal in the coal bunker. In the evening, I opened up the barber's shop and cut Jim, Gordon and George's hair.

8 August

Wind west-north-west with gales all day – mean speed 42 knots. Highest hourly mean 63 knots. Highest gusts 115 knots. There were many other gusts that were over 100 knots. Temperature -6°C (21.2°F) to -20°C (-4°F). Heavy drifts and snow with the visibility down to between two and ten yards (1–9 m) in the morning. Lowest pressure 967 mbs. For one met observation, Alan was roped up to get the screen to read the temperatures. Later though, we decided that the conditions were too dangerous, so two observations were done from inside. As it was Saturday, there was the usual weekend clean-up throughout the hut. When washing the floors, the water froze, making everything very slippery. In a slight lull in the weather in the afternoon, Jim and I managed to get to Observation Bluff for an ice observation. We saw the first clear water from Gourlay – Polynesia Point, Outer Islet, Sunshine Glacier (Coronation Island). The first open water seen for several months. Gordon had another failure with bread baking and Brian came to his rescue and baked for him.

10 August

Wind west-north-west strong to gale slowly moderating during the day with a mean speed of 27 knots. Temperature -15°C (5°F) to -18°C (-0.4°F) all day. The weather was fine. Lowest pressure 980 mbs. On early met duty, I did a check on sunshine statistic. Took usual temperatures around the hut and I also measured the snowdrift around the hut. Temperature down to 10s in living room and bedroom. Gordon borrowed my negatives and had a printing session all afternoon and evening.

In the last eleven days, on ten of them we had gale-force winds!

By the end of August, our first gentoo penguins were seen walking over the ice in the Gourlay area – they nested only at North Point. On 22 August, Brian and I saw seventeen Weddells with four pups only about a day old on the sea ice near Jebsen Rocks. That day, three of us had walked all the way around the island on the ice, looking for our first pups. At one place I went through a piece of rotten sea ice, but I managed to roll out, only getting my legs and feet damp. Also, we saw over fifty gentoos on the ice shelf at Gourlay, which I expect were waiting for the ice to break up so they could get to North Point.

A few days later, a further attempt was made to get to the west coast to check on the Weddells only to be forced back by strong winds. By the beginning of September, the ice edge had crept a bit nearer, but we could still get to Coronation Island. The next time we visited the west coast, there were 120 Weddells and seventy pups and still no sign of the ice breaking up on that side of the island. There were also a further thirty-three Weddells and nineteen pups at Gourlay. The beginning of September also saw the first elephant seals returning to Elephant Flats, along with a lot of bird life. There were sheathbills, Cape pigeons, giant petrels, gulls, skuas, and blue-eyed shags.

On 17 September we did another trip around the island on the ice; we counted 300 Weddells with 200 pups. Also, we counted 300 gentoo penguins at Gourlay still waiting for the ice to break up so that they could get back to North Point.

About this time, Fergus, who was in South Georgia, had several radio skeds with us. He wanted us to tag all the Weddell pups, but we could not as we had not been supplied with tags when the stores were unloaded the previous summer. Brian suggested branding the pups, so we sought permission to do this. We were granted permission as Brian had had branding experience when farming on Pebble Island. The only thing was that we did not have branding irons either, but George soon set to it and made a full set in the workshops – all with wooden handles too. However, by the time he had made the irons the sea ice had broken up, and we were unable to get to the pups.

Finally, by the end of September, the ice started to break up in the Normanna Strait and all down the west coast.

At the beginning of October there were 500 Adélie and 400 gentoo penguins at North Point, and the shags and giant petrels had started nest building. By the middle of the month, the elephant seals were back in strength in the Wallows and Elephant Flats, and we saw the first pups. The ice was finally breaking up in the bay, but there was pack ice everywhere, making it very difficult to take a boat out. Even so, we made a few trips across Pipeline Beach by boat, and on one of these occasions we saw and captured a giant jellyfish which was two feet (60 cm) in diameter. We brought back the jellyfish and preserved it in chemicals for Fergus.

The end of October saw the first Adélie eggs at North Point and the return of the first chinstrap penguin. We now spent several days visiting Gourlay, checking on Adélie flipper bands and scrabbling up and down Cape Pigeon Gully checking on Cape-pigeon rings.

Fergus came up with another bright idea from South Georgia – could we brand as many elephant-seal pups as possible in the Elephant Flats area? These biologists were always very good at asking other people to do their work! For example, Lance Tickell asked us to weigh prion chicks for him every morning and evening the previous summer. Anyway, George, Brian and Alan made several trips and managed to brand several dozen pups. The main problem they had was getting the brazier to heat up the iron enough. They took the pups a good distance away from the parents, branded them and then returned them. The pups were taken away so that the branders were safe from being attacked by the parents. About this time, Jim and I visited Three Lakes Valley to see if it was possible to get any samples from the lakes, but we found that the ice was still five feet thick (152 cm)!

One day in October, Gordon went for a walk, and someone saw him walking up the Orwell Glacier. Jim and I hurriedly roped up to find him, only to discover him wandering along Moraine Valley saying that he did not realise that there were crevasses on that glacier.

During early November, we heard that the *Shackleton* was coming to us first this season as all the other bases were still ice-bound. Panic! We had not expected a visit until about Christmas, by the previous schedule. Everyone started writing reports and letters.

One evening on Stanley Radio there was quite a long talk about all the giant petrels we had ringed on Signy Island the previous summer and where they had been recovered all around the world. This made us feel like our efforts were not all in vain – we had produced some results and had put ourselves and Signy on the map.

On 17 November the *Shackleton* arrived captained this year by Commander Turnbull. He, along with some of his officers, came ashore to inspect our base and to ask all about our winter. Our isolation was over. I think we all felt sad in a way; we all had got on really well with each other – even with Brian's singing! So, it was farewell to Gordon, George, and Alan – they were all bound for another base for their second year. It was also welcome to Ron Pinder (wireless operator) and Bill Mitchell (diesel mechanic), but we were given no replacement for Alan. Now we were down to three meteorologists with a busy bird-ringing season coming up. Also onboard the *Shackleton* was Mr Canning (Chief Meteorological Officer for the Falkland Islands and the Antarctic) coming around to inspect all meteorology stations. We were invited on board for a meal and for a film showing of *The Cockleshell Heroes*.

Summer, November 1958 to April 1959

A rather hectic week passed after the *Shackleton* departure. There were now only five of us left for the summer. We had to move boxes of food and other stores into the hut and check them, stack sacks of coals near the door of the hut, and roll drums of fuel within reach of the generator room.

Ron became our 'food minister', taking over from Gordon, who had departed for Horseshoe Island. Bill became the dog man, replacing George. I took over all the bird work from Alan, who was bound for Admiralty Bay. I was greatly assisted in this work by my four companions, either by helping with the bird ringing, recoveries and counting birds or by taking over some of the base work. Jim was often the only member to be left at base all day long doing the meteorology work, wireless operating and cooking while the rest of us were on the west coast ringing giant petrels.

During the summer we had to make a complete stock-take of everything we had on base and make a list of everything we needed to be delivered. On this list we put down everything from valves for radios, firebricks for ovens, light bulbs, paint, thermometers, fuel . . . and food.

One of the first jobs after we had sorted all the stores out was to collect several hundred chinstrap eggs to pack into boxes of flour for consumption later. The eggs were collected by boat from Gourlay and North Point. We did not collect from colonies where the oldest ringed chinstraps nested as we did not want to disturb them.

One day, early in December, we made a visit to Shagnasty Islet, off the south coast of Signy, to count the penguin population and we found a chinstrap and macaroni penguin together at a nest with one egg. They were obviously paired up together as they went into an ecstatic display, weaving their heads from side to side and braying.

We took several photographs of the pair while they were displaying. The thing was if the egg hatched what would the offspring be – a macstrap or a chinroni? During December we made several attempts to revisit Shagnasty Islet to check on the pair, but we were turned back by heavy seas and strong winds. Once we had to leave the boat at Gourlay and walk back. On another occasion, Jim and I thought we would try and cross to the islet, but the tide did not go out far enough for us to cross the rocks to the islet. We decided to count the penguins on the mainland opposite the islet and attempt to estimate the population on the islet using binoculars. Unfortunately, we were never able to get back to check on the pair.

The IGY programme of bird ringing was due to end on 31 December, so prior to that date we put a great effort into catching as many skuas as possible in the trap in front of the hut. Several times, I got up in the morning to find a note from Brian which said that he had sprung the trap early in the morning before going to bed after his night meteorology duty and that there was a skua or two in the trap waiting for me to ring. This is not to say that we didn't carry on ringing after 31 December. It was just that after that date we did not use the distinctive blue-and-white plastic band which indicated to anyone who saw the bird that it had been ringed on Signy Island during the IGY.

Early in December, Ron and I went on a seal-counting trip around the Stygen Cove and Elephant Flats area. We counted over 900 elephant seals. We also checked on the pups that we had branded earlier – all were fit and healthy. While in that area, we counted all the gull and terns' nests, measuring the size of the eggs and (in the case of the terns) mapping their nests. The gulls nested in one colony as protection from the skuas taking their eggs or young. On another occasion, the terns were mapped in Moraine Valley and around Gourlay. One day while visiting Gourlay we made thirty-six chinstrap recoveries, including two Laws and Sladen had ringed in 1949, when the base had first been established. The birds had been breeding then, so they had now been breeding for at least ten years. Chinstraps come back to breed with the same partners on the same nest site each year.

At North Point we had one pair of macaroni penguins in a chinstrap rookery in the same place as they had been the previous year, with one egg. Later in January we spotted that the egg had hatched and that they were raising a chick.

On one occasion, after a fairly heavy snowfall going down to Cape Pigeon Gully, to check on the rings, we found nearly all the Cape pigeons snowed in on their nests with just their heads showing above the snow. Ron and I spent several days in better weather conditions going up and down the gully making recoveries and ringing many Cape pigeons and several snow petrels.

Another job was marking a record of the sheathbill (Paddy) population at Gourlay. We made numbered markers from concrete which we placed by each nest, and we painted a rock by each nest yellow so that we could locate the nests easily. The sheathbills had made the nests in and under rocks around the penguins' rookeries, and they ate the penguins' eggs and, later, dead chicks.

When checking a skua's nest near the top of the Stone Chute, I got buzzed by the owners of the nest as I knelt to check on their chick. One of the parents took my woolly hat off my head with its talons and dropped it about a quarter of a mile away. Usually when near a skua's nest we held an ice axe or stick above our heads so that the skuas knocked that instead of us, but on this occasion I had laid my ice axe down on the ground beside me.

One day I accompanied Jim to Tioga Hill to make an ice observation. From there we went down over Gneiss Hills to overlook Confusion Point. We then descended to Pandemonium Point to count the birds in the chinstrap rookery there. Then we climbed up again overlooking Cummings Cove and went back to base via Tioga Hill – a good day's stroll!

The 6th of January was a day to remember. Bill, Brian and I boated to North Point to ring adult giant petrels and blue-eyed-shag chicks. We hauled our boat up over the high-water mark and tied it to a large rock. Then, while up on the cliffs ringing, we suddenly noticed our boat drifting out in the bay. It must have been a very high tide! We dashed down to the beach, but by then the boat was drifting out of the bay towards Coronation Island and nobody fancied a swim in the very cold sea. Gallant Bill volunteered to run back to base via the col between Robin and Jane Peaks, Three Lakes Valley and Elephant Flats to get another boat while Brian and I kept an eye on the disappearing boat. Back at base, Bill found the other outboard motor unserviceable, so Jim and Ron rowed all the way up to North Point to pick Brian and me up and hopefully to also collect the drifting boat. But by the time they arrived the drifting boat had disappeared towards Coronation Island amongst icebergs. Bill, in the meantime, worked on the unserviceable motor to get it working. The four of us rowed back to base, and by the time we arrived Bill had got the second outboard going. We quickly sorted out some supplies – sleeping bags and food, etc. – then we drew lots to see who would go searching for the boat! Brian and Bill drew the short straws, so they set off by boat towards the Cape Hanson area of Coronation Island – the direction the boat had last been seen drifting. Luck was with them, as they found the boat and towed it back to base after midnight – over eight hours after it had drifted out from North Point. As it was dark when they returned, we kept all the lights on in the hut so that they could navigate back.

Not to be put off, the next day Bill and I went by boat again via North Point to the west coast to ring birds all day. We took sandwiches to eat and lemonade crystals to make a drink using water from the mountain streams. That day we were more successful and managed to ring 108 adult giant petrels.

Again, the next day, Brian, Bill, Ron and I visited the same area. This time the sea was rough, so we only boated to Wallows and then walked via Spindrift Col to the west coast. I worked with Ron, and Brian and Bill worked together, and we ringed 210 adult giant petrels. We also recovered one adult giant petrel which Laws had first ringed near the base in 1948. There were no longer giant petrels near the base – they must have been driven away by dogs and man.

We made two more trips to the west coast and managed to ring a further 480 adult giant petrels. On one of these occasions, we did a seal count in the Wallows on the way back – there were 2,500 elephant seals on the beach.

We saw the odd fur seal around Berntsen Point during January and February. There were also about 1,000 elephant seals at Gourlay.

Late in January, Fergus came up on the radio of Deception Island. It was the usual biologist request – could we do this and that and collect some specimens for when he arrived in April. He wanted us to collect quite a few bird specimens, chloroforming them and then either preserving them in spirit or packing them in salt. However, we first had to weigh them and measure each specimen.

As a change from all the bird work that I was doing, I set up a meteorology screen on Garnet Hill equipped with a thermograph to get comparisons of the temperatures between the snow ice cap and base. This meant that I had to make weekly trips to the screen to change the chart. Also on some occasions I had to empty the screen of drifted snow. I remember once going up to Garnet Hill with Jim to visit the screen when the ice-cap surface was very icy, and we had to cut steps in the surface nearly all the way up. Then when we finally got there the low cloud enveloped us, and it began to snow heavily. We had a job finding our way back down even though we had marker posts every few yards all the way up there! We were also sending up balloons every time there was no low cloud to get readings of upper winds – especially when wave clouds formed as I was writing a paper about wave-cloud formations formed by the mountains on the western end of Coronation Island.

In mid-February we had a heavy snowfall, and we all got our skis out to go skiing in the middle of summer. It didn't last long though! A few days later, Ron and I ringed over 100 Cape pigeons in Cape Pigeon Gully, whereas a few days earlier they were almost buried up to their heads in the snow.

On 21 February, Bill and I set off for the west coast to camp and ring giant-petrel chicks. Brian and Ron took us by boat, and on the first day we put up our tent and established our base. Then on 22 February we ringed 400 chicks. The next day we were a bit stiff from all the previous day spent bending over and ringing birds, so we only managed to ring 200 chicks. The following day the weather was wet – water everywhere and low cloud on the hills! – so we had a day off and stayed in the tent. On the 25th we ringed a further 400 chicks. Everywhere was very damp as the thaw continued to cause streams of water to run everywhere. Also, all day and night we heard rumbles from bergs breaking up and avalanches crashing on Coronation Island. The next day we ringed another 180 chicks. We couldn't do any more as we had used up our complete supply of rings. Therefore the following day we walked back via Spindrift Col to base. Another day we made a boating trip to collect the tent and all our gear. That was the only camping trip we managed that summer due to the shortage of staff.

We spent another half-day after our ringing trip entering up all the details of the ringing in the register, and made a duplicate copy to send to the FIDS scientific bureau. By this time, we had also used up our last skua ring, so ringing came to a virtual stop, although we still kept up our regular seal counts all around the island. By March, the elephant seals had started to leave the island after their summer moulting.

Another task we had during February and March was to do with penguins. Early in March 1958 we had received a telegram from the FIDS secretary to ask if we could collect twenty-five Adélie and chinstrap penguin chicks to be picked up by Salveson's whale catcher for Edinburgh Zoo. We sent the reply, 'No can do. Gone Fishing.' All the Adélie's had left a week or two earlier at the end of the breeding season.

Eleven months later, on 12 February 1959, we had a radio sked with the administrative officer (AO) in South Georgia. He asked if we could try and catch and keep some Adélie and chinstrap penguins in captivity until Salveson's whale catcher could come and collect them, at the earliest on 10 March, at the end of the whaling season.

On 15 February, Brian, Bill, Jim and I went overland to Gourlay Peninsula and collected twenty-five Adélie chicks to bring back to base. We put them in sacks on a sledge pulled by Garth, one of our huskies. On the old whaling plan in front of the hut we had a large pup pen which we put the penguins into. I informed the AO in South Georgia that we had completed the first part of the mission and that we would leave collecting the chinstraps until just before the whale catcher visited as they did not migrate as early as the Adélies.

It was then Operation Food to feed the penguins. Brian and I boated up to Shallow Bay, where we shot a seal and towed it back to the bay. We cut up the seal and minced up some of the raw meat and mixed it with cod liver oil and water. Then we force-fed them on this mixture. They didn't seem to object too much to it and appeared to enjoy it. So we minced up a lot more seal meat to feed them for the next few days.

Unfortunately I cut my finger when cutting up seal meat to feed the penguins and my finger became swollen (seal finger). After a sked with the Doctor at Hope Bay, I treated it with antiseptic cream and kept my arm in a sling for a few days.

Two days later we visited Gourlay and found that all the Adélies had now left. It was a good job we got them when we did.

After a few days, the penguins were getting rather dirty and muddy, so Ron and I washed them down with a stirrup pump and then wiped them down with old rags and brought them into the workshop in the afternoon to dry off. I had another sked with the AO in South Georgia to let him know that everything was progressing all right. He asked if I could have some kind of document prepared for the whaling company to show where the penguins came from to produce for customs in England.

We thought the penguins could do with a change in diet, so Ron and I got the fish trap out, baited it with seal meat and sunk it in the bay. The first catch produced nine fish, which we filleted and fed in a long strip to the Adélies. Five of them were now taking food from the hand and were not having to be force-fed.

Between 21 and 27 February I was away from base, camping on the west coast of the island, ringing giant petrels (as detailed earlier), so I left the Adélies in Ron's capable hands. Upon returning, I found that Ron had caught forty-four fish in the trap. Therefore I helped him to gut, skin and fillet the fish.

By 8 March we were running short of food for all the penguins, so all but Jim walked to the Cemetery Flats, where we shot a bull elephant seal and carried the meat back in backpacks via the Stone Chute. Unfortunately, on the way back I ruptured myself carrying the seal meat. There was swelling on the right side of my groin although it was not painful at all. I had a radio sked with the Doctor at Hope Bay, who just told me to carry on as normal.

On 11 March the captain of the whale catcher contacted us and said that he was very grateful for all the work that we were doing with the penguins and that he would be arriving at Signy sometime the following week.

On 15 March, in the afternoon, Bill, Brian and I set out by boat to collect twenty-five chinstrap chicks from Gourlay. However, we had to turn back in Paal Harbour due to rough seas – the wind was between twenty and forty knots all day. The next day Bill, Jim and I again set out by boat in the morning, and this time we collected twenty-five chinstrap chicks. On the way back, we were followed by a leopard seal, which seemed to know that we had a good meal on board. He kept diving under the boat, first coming up on one side and then the other. In the evening, we heard that the *Southern Gem* was due the next day at 1 p.m.

The 17th of March was a sunny day – the best day we had had all summer! The *Southern Gem* arrived at 1 p.m. and anchored in the bay.

We safely boarded all the penguins in big cages made especially for them. The crew entertained us well on board, and five rather drunk Fids finally bade farewell to the penguins destined for Edinburgh Zoo. I had always thought that drink was banned from all whaling companies and ships – perhaps they made a special exception for us. The whaling company gave us a crate of oranges (I hadn't seen an orange for eighteen months), beer, a Norwegian sweater each, twenty-six whale teeth, lots of rope and canvass, and a bottle opener made from a whale's tooth. I handed over my home-made official document stating that the penguins were exported from the South Orkneys with the approval of the base leader at Signy Island (me!). The captain of the whale catcher asked me what the weather would be like for his trip to South Georgia, and looking into my crystal ball he decided that he would have a good trip with light winds (the pressure had been rising for several days and was high for that part of the world).

Two days later we had a sked with AO South Georgia. All the penguins had arrived safely at the whaling company, and they had caught a ton of fish to feed them on their journey to England on the whaling factory ship. The whale catcher's skipper sent his thanks for a good forecast – it was the best weather he had had all season. We passed on our thanks via the AO to the whaling skipper and company for all that they had given us.

On my return to England over a year later I visited Edinburgh Zoo and saw some of our Signy Island penguins.

At the end of March, we had a signal through from the FIDS secretary saying that due to bad ice conditions the *John Biscoe* had been unable to relieve some bases and therefore some staff were being redistributed. The Signy base was going to get three extra chaps as well as Fergus for the winter – how we could have done with them in the summer when we had all that ringing to do! The FIDS equipment officer was on board the *John Biscoe*, and he asked us over the radio what extra supplies we required as they had a surplus on board. It should have gone to Base W, but it had been closed due to the ice. Ron and I did a quick stock-take and calculation before putting our order in.

The beginning of April saw winter returning. The ice was now three to four inches thick on the lakes in Three Lakes Valley, and odd patches of sea ice had started to form in the bay.

By 6 April, the *John Biscoe* was still stuck off Graham Land in 10 tenths heavy pack ice. The American ice-breakers *Edisto* and *Northwind* were going to her assistance. Finally, on 20 April, the *John Biscoe* arrived at Signy with the next extra personnel, plus extra dogs, Fergus and the supplies. The doctor on board, Dr David Jones, inspected my hernia, which I had developed while carrying seal meat in a backpack. He had a radio sked with Dr Slessor in Port Stanley, and it was decided to ship me out for an operation. I had a very quick handover to Jim before I packed my bags and boarded the *John Biscoe*. I was sharing a cabin with Brian, who was going home after just one winter. Three days later, in South Georgia, I met Lance Tickell (the person we had to weigh prion chicks for our first summer), Mr Matthews (the administrative officer whom we had had radio skeds with concerning the Edinburgh Zoo penguins), and Gunner Johansen (manager-foreman of Leith and gunner on the *Southern Gem*, which collected the penguins for Edinburgh).

The next day we sailed for the Falklands. We heard on the radio of Alan Sharman's death – he had been found at the bottom of the cliffs at Admiralty Bay.

It was a very rough trip back to the Falklands. I did the daytime meteorology observations from the bridge. At one time the ship was rolling thirty-seven degrees in each direction. On 27 April we finally arrived in Port Stanley.

Part Three: The Falklands and South Georgia

Eight Weeks in the Falkland Islands

The *John Biscoe* arrived in Port Stanley on the afternoon of 27 April. I was back in civilisation again for a while. Dick Smith from the meteorological office met me at Port Stanley. One of my first tasks was to see John Green (the FIDS secretary) to sort out my future. We agreed that if Dr Slessor was willing to operate on me sometime in the winter, I was to spend the winter months up to next October–November in Port Stanley Meteorological Office preparing weather statistics and annual tables of records for the FIDS bases. That entailed weeks poring over figures and operating the adding machine, after first analysing the continuous twenty-four-hour wind-chart records for Hope Bay for a whole year. Hope Bay had not, unlike other bases, analysed their records hour by hour, day by day, month by month, as they were too busy with other scientific work, such as surveying on long sledging trips. Therefore there had only been few staff left on base to make the routine three-hourly meteorological reports and to complete all other base duties. After that it would be my task to prepare all the meteorological stores to be sent south on relief ships for the next summer season. Then the next summer it was intended for me to go south again as a biological assistant, most likely helping Fergus O'Gorman with his study of the southern fur seal.

One of the first events for all returning base leaders was to be invited to dinner with the Governor at Government House. Joe Lewis from the meteorological office came to my assistance by lending me a suit, as I hadn't brought one with me! Joe had been south in 1954 and 1955 in the time of Ellery Anderson at Hope Bay, and reference to Joe and photographs of him can be found in *Expedition South* by Ellery Anderson. The dinner was very interesting and informative as we exchanged stories of our experiences.

After dinner, we all sat around the fire in easy chairs reminising, drinking and going over the tragic loss of life. We mulled over how these accidents had happened and whether they could have been avoided. In a way, it seemed more significant to me than to some of the others as Alan Sharman and Geoff Stride had travelled down with me on the *Shackleton*. Geoff had spent part of the first summer on Signy until George White arrived, and Alan had been on Signy a full year with me as a meteorologist and bird man before going on to Admiralty Bay. The other fatalities – Dave Statham and Stan Black – had left Signy as I arrived, which meant that all four were ex-Signy-ites, although none were on Signy Island when tragedy struck. We finally bade our goodnights to the Governor at around 1 a.m.

The menu at Government House was grapefruit, soup, fish course, beef with fresh vegetables followed by pears and cream. It was a slight change to a base dinner of tinned soup, corned beef, dehydrated cabbage, and carrots with powdered potatoes followed by prunes and tinned milk.

A few days later, all the Fids off the *John Biscoe* and local government workers attended a cocktail party at Government House. Afterwards, many of us went off to various homes to continue the entertainment, drinking and swapping stories. I ended up at Joe and Jean Lewis's home, at the top of the hill overlooking the harbour, along with Jim Shirtcliffe (another ex-Fid who had been to Signy Island, amongst other bases) and several other Fids and meteorological-office people.

All of the time the *John Biscoe* was in Port Stanley I stayed on board, but I had to start looking for somewhere else to stay. Many of the meteorological-office staff stayed at The Ship Hotel. I spent several evenings with Brian Beck at Eric and Eileen Ward's (Eric was a radio and radar technician at the meteorology radio station), and they invited me to stay with them. I much preferred this idea to staying at The Ship Hotel.

Eric was rather a madcap. He played the bagpipes (Eileen used to banish him to the peat shed or the peat bog to play), he was keen on loud band music and had many records of pipes and drums. He was also a keen amateur-radio ham and used to call a friend in England each week to chat about the weather, sport, etc. His radio equipment was set up in the large kitchen, so he could sit in the warmth near the Rayburn while making his contacts. One of his contacts came in useful to me later.

Before starting work at the meteorological office, I helped Eric Salmon in the FIDS office for a week, sorting out various things, and spent several pleasant evenings at his home with Frieda, his wife, enjoying meals and chats.

Finally, on 8 May, the *John Biscoe* departed for England. I moved into Eric and Eileen's house and started working for the meteorological office. It was a rather relaxing routine. I spent mornings in the office working on the weather data. Then I went down to The Ship Hotel at lunchtime for a couple of pints of beer with the Radio Sonde crew and any other meteorology people who were around. Next I would have lunch before going back to the office for another hour or two before going home. I would often spend the evenings at the working men's club playing table tennis or darts in the meteorolgical-office darts team at various pubs around town. It was funny how we nearly always lost the darts match, but won the beer leg after.

About this time, Peter Hale, whom I had known at Dunstable Met Office, arrived to replace Dick Smith. Also, Mike Hulbert turned up, whom I had first met when I joined the office at Tangmere in 1951. His father had also been my maths teacher at Chichester High School.

I along with the usual meteorology crowd went to the May Ball in the Town Hall on 25 May. Robina from the FIDS office was the May Queen that year. Her mother, Margo, was a rather eccentric hairdresser who ran the Falkland Islands Company hairdressing salon. I seem to have taken Elaine (meteorological-office secretary) to the dance, but then accompanied Evie Mckay back to her digs.

On 3 June the senior meteorological officer, Sam Glassey, called me into his office and asked how would I like to go to South Georgia in just over two weeks' time if I was operated on straight away? I said, "Yes, please!" So I had to go along to the Colonial Secretary's office to meet Denton-Thompson, the Colonial Secretary. He explained to me that they were a meteorologist down in South Georgia due to someone backing out at the last minute. The FIDS allowed me to go until the summer, when they expected new recruits. The Falkland Islands Government operated meteorological-office work in South Georgia whereas the FIDS operated the Antarctic bases. Denton-Thompson and I had a long chat and I seemed to impress him. We had a good understanding, which was of great help later when we had issues with Captain Coleman in South Georgia. Denton-Thompson only had one arm, yet he used to drive a Land Rover around Port Stanley. He used to hold the steering wheel steady with his body while he changed gear with his hand. The RMS *Darwin* made a once-yearly trip to South Georgia, in the middle of the winter, and it was intended that I should sail on her. Two wireless operators were sailing down with me; they had been given a crash course in meteorology work in Port Stanley. If not, I would have been the sole meteorologist on South Georgia.

The next day I entered the hospital, where I had a small room all to myself. At the time the hospital had fifteen patients and a staff of two doctors, one matron, one sister and four nurses. I entered the operating theatre at 10 a.m. on the 5 June. Both Dr Slessor and Dr Ashmore were present at the operation, as well as the sister and Evie (the nurse). I was given a spinal injection so that I was numb from the waist down, and a sedative that should have sent me to sleep – but it didn't work! I saw most of the operation reflected in the metal reflector around the light until Dr Ashmore saw that I was looking and got the sister to cover my face with a sheet. I remember they had difficulty fitting me on the operating table as my feet stuck out over the end. The same happened when they took me back to my room – either my head was against the rails at the head of the bed or my feet were stuck out over the other end. I couldn't help because I was still numb from the injection. I just enjoyed their difficulties as they tried to fit me in! Evie seemed to have a soft spot for me as she was in and out all day to see how I was. Then she changed her duties, so she was on night duties for the rest of my stay. Elaine came most evenings to bring me supplies of cake that she had made. This is the life! Anyway, my operation came just in time for me to miss a big party at The Ship Hotel. The day after, I heard all about it as Brian Waudby, Glyn Pugh, Dave Bolt, Mike Burns, Eric Ward, Elaine Reeve and Jean Lewis visited me – all seven of them at one time! The Matron had let them all in!

I was not allowed out of bed officially till 16 June, but several nights before I had been out and around the hospital at night with Evie as my guide, visiting the kitchen for food and cups of tea!

One of Eric's amateur-radio friends kept my parents informed of my trip to the hospital, my progress and later my departure to South Georgia. Eric told him of my movements and then they wrote to my parents. It was all very unofficial, as amateur radio operators were not supposed to pass personal messages over the airwaves. If either of them had been monitored or caught, they could have lost their licence to operate. I was very grateful, as were my parents, as communication by official channels was very irregular. Mail only went in and out of the Falklands once a month, and then it travelled by sea via Montevideo in Uruguay.

On 17 June I was out of the hospital one day after first being allowed out of bed, and I went back to Eric's with instructions that I must not do any heavy lifting for several months. During the time I was in hospital, Eric and Eileen had adopted a baby born illegitimately to a local girl. Eric unofficially called him 'Thumper' because of the noise he made – they had been unable to have children up until then. When I returned via the Falklands the following year, they had produced twins on their own.

On 18 June I attended a cocktail party at Government House given by Denton-Thompson and met Captain Coleman, who was to be the new administrative officer in charge at South Georgia, and Mr Ruddy, the new customs officer. They were travelling with me on RMS *Darwin* along with Ronald Carter and Ronald Houlton (the two Rons), the two wireless operators.

The night before we sailed, I went to the fancy-dress dance at the Town Hall with Elaine; but I couldn't dance much as both doctors from the hospital plus the Matron were there, and I don't think they would have approved. On the 20th I boarded the RMS *Darwin* for South Georgia, and we sailed at 2.30 p.m. Many of the meteorological-office staff saw me off plus Elaine, Eric and Eileen.

The journey south-eastwards was rough. RMS *Darwin* weighed just under 1,800 tons, and she was made for working in shallow waters. She could sit on the seabed at low tide for loading and unloading, so she had a fairly rounded bottom, which meant that she rolled easily in heavy seas. The weather got colder each day, and by the third day we saw our first icebergs. The skies were busy with many Cape pigeons skimming over the waves along with giant petrels and albatrosses. On the fourth day after leaving the Falklands, we entered Cumberland Bay and tied up to the jetty at King Edward Point – we had arrived in South Georgia!

Shackleton's grave in the whalers' cemetery.

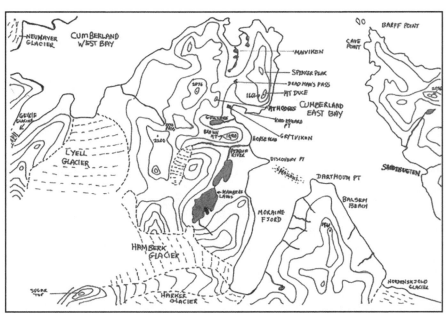

Local area round King Edward Point.

Wandering albatross, Bird Island.

Grey-headed albatross colony on Bird Island.

Fur seals on Bird Island.

Grytviken whalers' church.

The whaling station in full production.

Fin whale on the plan.

Otto Larsen on the harpoon gun on the Catcher 1.

The C. A. Larsen (tow boat) bringing in six whales.

Flowering mosses, Maiviken.

Lyell Glacier – lunch break.

Cave at Maiviken.

Echo Pass on the way to West Cumberlam Bay.

The face of the Lyell Glacier.

King Pengiun, Lyell Glacier

The John Biscoe in Port Stanley

South Georgia

South Georgia is a long and narrow mountainous island, just over 100 miles long with a breadth varying from three to twenty-five miles on an axis lying north-west to south-east. It has peaks 6,000 to 9,000 feet high (1,828–2,743 m) separated by deep glaciated valleys. The long north-eastern coast has many fjords which penetrate deep into the mountain range. The largest of these fjords is Cumberland Bay, which lies just beneath the highest peak, Mount Paget (9,625 feet). In the bay is King Edward Point. All the whaling stations were along this north-eastern coast, which is well protected by high ground from the predominantly westerly winds and weather. The south-western coast of the island is much more rugged, with steep mountains rising from the sea and none of the well-sheltered anchorages that are found on the other side of the island.

The first recorded sightings of South Georgia were just before 1700, but the first landing and exploration were not until 1775, when Captain Cook arrived. On returning home, his subsequent report on the abundance of fur seals came to the attention of the sealing industry. By the late 1700s, boats were returning fully laden with fur-seal pelts. In one season they took 112,000 fur-seal skins. At that rate, sealers soon depleted the fur-seal stocks; so the sealers turned their attention to the elephant seal, which they killed just for the oil from the layer of blubber under their skin. This oil was very high-grade oil. They left the rest of the seal's carcass for the giant petrels and other scavenger birds to clean. We could still find old try-pots in many of the bays around the island as a reminder of the slaughter. Later, legislation was brought in to protect all the remaining fur seals and to save a similar slaughter of all the elephant seals. Sealers could now only kill bull elephant seals over three metres in length. They divided the island into four quarters, and in one quarter each year no sealing was carried out. When I was there, Nigel Bonner, the sealing inspector and biologist, accompanied the sealers to monitor their work and to further his research into the seal population. They restricted the sealing season to a short period in the spring, before the breeding season and the opening of whaling.

The Argentine whaling company Pesca established the first land whaling station at Grytviken in 1904. Soon there were more whaling stations established, and at the height of whaling, around 1918, there were seven land stations on South Georgia. These were at Grytviken, Leith, Stromness, Husvik, Godthul, Prince Olav Harbour and Ocean Harbour. When I was there, in the summer of 1959/60, there were only three stations in operation, at Grytviken (manned mainly by Norwegians operating the *R* whale catchers), Husvik (manning the *Busan* whale catchers) and Leith (operated by a Scottish company with a mixture of Scottish and Norwegian crew members on their *Southern* whale catchers). Stromness was also open, but only as a maintenance and repair station. It had its own floating dock for the repair of Leith whale catchers and for the factory ships, such as the *Southern Harvester* and *Southern Adventurer*, which roamed the Weddell Sea in search of their catch. It was in the floating dock at Stromness that they temporarily repaired the RMS *Shackleton* after her accident in 1957, just after she had landed me at Signy Island. The manager of Pesca was Mr Ringdal, and I had to ask him for permission to go on any of the Grytviken boats. The winter manager at Leith was the captain of the *Southern Gem*, which visited Signy to collect the penguins for Edinburgh Zoo in March 1959.

Unlike on Signy Island, where there were only six of us, at King Edward Point there was a small community of about twenty personnel, including wives of five of the workers. We were also in constant contact with the whaling companies and their workers. Ships, although they were very irregular, were busy in the summer taking whaling personnel in and out. There were also many tankers shipping out whale oil. However, in the winter other than the RMS *Darwin*'s visit there were no visiting boats. So in the summer there were plenty of irregular opportunities for mail to come in and out.

The magistrate, Captain Coleman, was the officer in charge. He was responsible for overseeing the administration of the whole island, including the whaling companies. Under him was Mr Ruddy (customs officer), who had to clear all the boats entering and leaving the island. Each person on the point did his own job and was not expected to muck in and work as a team helping the others, as we did on Signy. There were a cook and steward (Nancy and George Mowatt) to provide meals for all the single and unattached men. In my time at King Edward Point the staff was

Magistrates	Captain Coleman and his wife
Customs Officer	Mr Hillary Ruddy
Chief Radio Officers	Peter McCleod and Pearl McCleod
Radio Operators	Ronald Carter and Ronald Houlton
Diesel Mechanics	John Quigley and Eric Brumby
Constables	Basil and Betty Biggs, and Jock Lawrence
Magistrates' Cook/Steward	Jimmy Shields
Biologist / Sealing Inspectors	Nigel and Jennifer Bonner
Meteorological Forecaster	Danny Borland
Meteorological Observers	Michael Meade, Dave O'Regan and me.
Painter and Decorator	Bill
Cooks and Stewards	Nancy and George Mowatt

As on Signy Island, in my time in South Georgia I joined forces with Peter McLeod, and between us we became the barbers for the point, often cutting four or five people's hair at one session, either in the table-tennis room or in the meteorological office.

During the winter we had two whaling personnel from Grytviken coming over by motor boat each day to do internal decorating of some of the buildings. One of these, Rolfe Olsen, I got to know well. His home was Sandefjord in Southern Norway, which was home to a large percentage of the whaling fleet in the Antarctic winter.

Buildings on the point were a customs shed/post office, Discovery House, the magistrate's house, the customs office house, the handyman's house, a radio station/house and games room, two bungalows (these were taken down on the *Shackleton* in 1957), the meteorological office, the jail, and a boat shed/store.

In the games room there was a full-sized snooker table, a table-tennis table and a dartboard.

Another thing that was different from Signy was that they did not provide us with all the extras, such as soap, toothbrush, and toothpaste, chocolate, camera films, boot polish, clothing, etc. Instead, we could buy these items at the whaling station 'slop chest', which was run by Mareceleno Jensen. The food was provided free, and I still had all my Antarctic clothing allowance.

Eric Brumby and I helped with various tasks other than our own. I mostly helped Niger Bonner, the biologist, when I could. It made life more interesting and gave us the opportunity to get out and about once the summer arrived. The rest of the personnel stuck rigidly to whatever their job was.

They first started making meteorological records in South Georgia at King Edward Point in 1907, and they continued to do this up until the Argentine invasion in 1982. The climate of South Georgia is subantarctic as it is 150–200 miles south of the Antarctic Convergence and relatively near to the Antarctic continent. It is kept cool by the sea currents circulating out of the Weddell Sea laden with pack ice and bergs. We experienced sub-zero temperatures throughout the year, but it was not as cold and inhospitable as Signy. The weather that we recorded at King Edward Point was not representative of the rest of the island because the mountains shelter it to the south and west. The föhn winds blowing down the lee side of the mountains often warmed King Edward Point, when the temperatures could quickly rise by 10°F and the temperature could reach 21°C (70°F) in the summer. A comparison of the weather at King Edward Point and at Bird Island (off the north-west tip of South Georgia) shows a remarkable difference in temperature. Over a period when biologists were making weather records on Bird Island, the records showed a maximum temperature on Bird Island of 9°C (48°F), whereas they recorded 22°C (71°F) at King Edward Point. Similarly, it was much sunnier at the point and a lot less windy, even though we did not receive any sunshine for several weeks during the winter due to living in the lee of high ground. All the whaling stations also made weather records at various times.

In the winter, South Georgia is completely snow-covered, but in the summer the snow line is about 1,000 feet (304 m) above sea level on the north-east coast and very near sea level on the south-west coast.

The average temperature for King Edward Point was 2°C (36°F), and for three months in the winter the average was below 0°C (32°F). Snow can occur at any time of year.

My First Three Winter Months
at King Edward Point

I arrived at a snow-covered King Edward Point on 24 June, when there was a thin coating of slushy ice on the sea in part of Cumberland Bay. The two resident meteorologists, Jack Ford and Joe Cochran, were both leaving on the RMS *Darwin* for home and were only too pleased to see me. After the briefest of handovers, the meteorological office at the point was all mine – except for a little help from the two Rons, who were going to do two observations on alternate nights until Danny Borland and two observers arrived at the beginning of the whaling season. This meant that I was on duty from 7 a.m. to 10 p.m. every day.

My job was to run the meteorological office and make weather observations every three hours, compile weather statistics, send up balloons when the weather permitted to ascertain the upper wind direction and speed, and to maintain all the instruments.

Later, all throughout the summer, when Danny was there forecasting for the whaling industry, I helped him in taking the Morse messages of the weather reports from the various stations, including the Falklands and Antarctica (FICOL), Argentina (LSV), Chile (CCS and CMM), Brazil and the rest of South America (PMM), and South Africa (CFX). Then I would plot the weather charts of these reports for Danny to base his forecasts on. South Africa used to collect then pass all whaling ships' reports with their positions coded. Each whaling position had its own position code for its whale catchers, so other companies couldn't intercept their broadcasts to locate where they were hunting. We had a position decode for the South African broadcast which we were forbidden to show to any visiting whalers. Whaling was a very secretive business – you never let any other companies know where you were. Taking Morse code was a bit confusing at first until I got used to it. Some countries sent out five dashes for 0; others used three dashes and still others just one long dash. Danny had been taking Morse for so many years that he could sit taking Morse and do a crossword at the same time.

To make upper-air balloon ascents, we used balloons that when filled with hydrogen ascended at 700–1,000 feet a minute, depending on how much hydrogen we used. The balloon was inflated with hydrogen and balanced against a weight. When the balloon floated without rising or descending, we knew that it was time to remove the weight, and the balloon would then ascend at the given speed corresponding to the weight used. Next, after orientating and levelling the theodolite, we would release the balloon. Then we would take readings every minute of the elevation and bearing of the balloon. After the balloon had either disappeared behind a cloud or a mountain, or burst, or we had got too cold, or it had got too small and lost, we would analyse the readings with the help of a special slide rule to obtain the relevant winds. We coded these results and sent them to the meteorological office in Port Stanley, who broadcast them to anyone who was interested along with the next weather report. We generated the hydrogen ourselves with a more powerful generator than the low-pressure one at Signy, which made just enough hydrogen for one balloon ascent.

I stayed, as the sole occupant, in one of the new bungalows that we had delivered in November 1957. The bungalow was located at the opposite end of the point to the meteorological office, not far from the cross which members of *Quest*'s crew had erected when Sir Ernest Shackleton died in 1922. I had quite a walk to work in the snow of winter! I had all my meals (prepared by Nancy and George Mowatt) in Discovery House along with all the other single or unaccompanied workers.

By the second week in July, the snow had become too deep to walk in, so I skied back and forth between the bungalow, meteorological office, radio station and Discovery House. One of the tasks we had to do with each meteorological observation was to make a tide-pole reading at the end of the jetty. This was a bit hazardous at night-time as I had to ski out on to the jetty, surrounded by icy cold water, while holding a torch in one hand so that I could see the readings on the pole. Still, I never fell in!

In time, the snow got so deep that I was skiing over the washing lines and I had to dig out the meteorology screen as the snow covered the base of the screen. Each day I had to dig out the doorways to both my bungalow and the office.

On 12 July, between weather observations, I along with several others from the point skied around the bay to Pesca Whaling Station to watch the ski-jumping competition of the wintering staff. The competition was won by Harry, who had been a steward the previous winter at the point.

But then, whilst Captain Coleman and his wife were visiting Husvik Whaling Station, someone broke into his house on the night of 18/19 July and took several bottles of gin. Immediately, suspicion fell on the same Harry who had just won the ski-jumping competition a few days previously. He had been in trouble only three months earlier for drunkenness. All whaling stations were supposed to be drink-free, and one could only obtain alcohol from the point. When Captain Coleman returned on 19 July, Harry was arrested and charged with the break-in, and his trial was set for 22 July. The court was to be held in the customs building by the jetty, and Mr Ruddy was to act as the accusing council, but who was to act as magistrate? Captain Coleman was the magistrate, but it was his property which had been burgled, so he could hardly be impartial. I was almost clobbered for the job as I had been sworn in as magistrate for Signy only eighteen months before in front of the Governor in the Falklands. But somehow the rules got bent after a radio sked between the Governor and Captain Coleman, and he sat in judgment. On the day of the trial, I was between observations, so I went along to attend the trial. Harry's English was not too good, so he had an interpreter to help him. He pleaded guilty, and Coleman imposed a four-month imprisonment on him with hard labour, which caused a gasp of astonishment from those attending as they thought it was a harsh sentence.

Harry spent the next four months locked up in the small jail. Each day he did supervised chores around the point, being overseen either by Jock Lawrence or Basil Biggs, our two constables-cum-handymen. While sober, Harry was a very amiable chap – he just had a drinks problem. They released Harry in November.

By the end of July, the weather had turned very cold, and the water supplies were becoming very low – nearly all the water in the dam holding our water supply had frozen up. The bungalow's supply was also completely frozen up, even with the heating inside the building. I ended up cutting snow blocks, as I did at Signy the previous winter, for the water tank so that I could wash and use the toilet. That July turned out to be the coldest since people started doing full records in 1945.

On 31 July I heard on the radio the sad news that Keith Bell, who had travelled down with me on the *Shackleton*, had fallen down a crevasse and was killed at Base F.

The 1st of August saw a party of whalers from Pesca go on a deer hunt. They sailed on the *Petrel*, an old whale catcher, across Cumberland East Bay to land leading out to Barff Point. Before departure, permission for the deer shoot had to be granted by Captain Colman, and a record kept of deer shot. The deer had been introduced from Norway many years before to provide sport and food. Five deer were shot, and a few days later we had venison on the menu.

August saw more heavy falls of snow. On the 12th eighteen inches fell in three hours, making travelling very difficult. Later, with very low temperatures, the snow developed a good crust on top and skiing became good and fast. We used to ski down the slope by Shackleton's Cross and out over the flat area which was a football pitch in the summer, where later the new accommodation building, Shackleton House, was built. One day I skied over to Pesca in just seven minutes over the icy surface. The trick to do the journey quickly was to follow in the tracks made by someone before you – as long as they had cut good straight tracks which had later developed a frozen crust, it was like skating along. You were not having to form your own tracks through virgin snow, which needed a lot more effort.

On 20 August we had the heaviest snowfall that could be remembered by any of the oldest inhabitants. There were even avalanches all the way along the side of Mount Duce from Shackleton's Cross to Pesca. The avalanches came very near the buildings and completely closed the route to Pesca. At this time, quite a lot of pack ice drifted into the bay with many seals on it. The lowest winter temperature was -13°C (9°F), which we recorded on 30 August. It couldn't be long until winter left and spring returned!

On the last day of August, the first boat to visit the island since my arrival was the *Conquistador*, an Argentine tanker with the first of the summer staff for the whaling station, plus fuel for the catchers. However, there were no meteorologists on board! I had been told that there might be someone coming, and at this news the two Rons had decided to stop doing their two nightly observations. I was on my own! So I continued doing my 7 a.m. – 10 p.m. day plus making up the two missing night observations from the recording instruments (thermograph, hygrograph, barograph, and anemometer).

Throughout the winter, for indoor entertainment we had the snooker, table tennis and darts in the room off the radio station. Peter McLeod and I had an ongoing tournament with the other two radio operators – it was a case of the two Peters versus the two Rons. We used to have the odd game in the afternoon, after I'd sent the meteorology observation, provided there wasn't too much radio traffic. I think that at the end of the winter Peter and I were the winners. Others used to play cards in the Discovery House lounge. There were also the odd parties at either the Colemans', Bonners', McLeods' or Mowatts' houses, or at John Quigley's room in Discovery House. There, he used to serve up rum punch made from draught rum, orange juice, a little sugar and boiling water. I think his punches were about seventy-five per cent rum – they were certainly strong and had a kick! Outdoors, skiing was very popular when the conditions were good. During the winter, people at Leith held a winter sports meeting, where most of the people from the point went. They were taken there on the old whale catcher, the *Petrel*, from Pesca. I couldn't go, being the only meteorologist, so I cooked the meals for the ones left in Discovery House. I heard afterwards that the winter manager was asking where I was – he was the gunner on the *Southern Gem*.

Early in September the thaw set in, and there was slushy snow everywhere. We now needed wellington boots instead of skis and ski boots. On 16 September the walking conditions were at their worst – I was sinking up to my knees in snow and slush.

It was about this time that I heard that John George, who was supposed to be my replacement, was not coming, so I wouldn't be able to go back further south in the summer. Instead I would have to remain in South Georgia. I called him a few names under my breath as I had been looking forward to seeing more of the Antarctic. Still, two more observers were due at the end of the month plus Danny Borland, the forecaster.

The beginning of September also saw the start of the elephant sealing, so Nigel Bonner was off around the island on the sealer boats monitoring the slaughtering. He also made seal counts and tagged pups. That year Cumberland Bay was one of the areas where the authorities had permitted sealing, and I remember seeing one of the sealing boats lying offshore across the bay by Hestesletten Plain. The snow on the beach was red with blood – they had left the carcass to rot after they had removed the blubber and towed it out to the waiting sealing boat. Various birds soon surrounded the carcass, and in only a few days only the skeleton was left – plus, a lot of fat overfed giant petrels, gulls and skuas. In three days, one sealer skipper, Hauger, on board the *Albatross* collected 150 elephant-seal skins and blubber. The hold was full, and the decks, when he returned to Pesca for the blubber to be rendered down. The quota for the season was 6,000 elephant seals.

The evening of 14 September was a lovely clear night, so I looked at the moon through the theodolite at the time the first Russian space rocket was due to crashland on the moon's surface. I don't know what I hoped to see, although it was a brilliant night for moon-gazing.

At the end of the month, finally, Danny Borland and the other two meteorologist observers arrived on the *Opal* at Leith. The next day they arrived by whale catcher at the point. The two new observers were Michael Meade and Dave O'Regan from the Irish meteorological service. This was the first time anyone from the Irish meteorological service had ever been abroad. I quickly introduced them to the office, showing them around and explaining what to do. Neither of them could receive Morse, so it was left to Danny and me to do all the receiving. Then on 1 October they started work – we now worked two days on, followed by two nights, and then two days off. Danny was around during the day to do the whaling forecast. At this time, all the whaling boats were coming in from Norway and Scotland, so the harbour was very busy. At least now I could have some time off and would be able to get around the island a bit, helping Nigel Bonner with his biological work – or so I thought (more on that later). I also visited Arthur, the South African doctor at Pesca, and he passed me fit to continue active work.

Summer In South Georgia

The summer weather could be very changeable. One day we could be having gale-force winds and snow showers. Then a couple of days later there would be blue skies and föhn winds with temperatures reaching 21°C (70°F). By mid to late October the snow had melted enough around the shores to walk around the coast, but inland over Hestesletten Plain it was still too deep and slushy to walk. By mid-November, all the low-lying snow had gone and walking became easier, although there were running streams of melted snow everywhere.

Coming south, I thought that I would get away from summer attacks of hay fever, but the flowering of tussock grass meant that at times in the summer my eyes were running and my nose streaming.

With the end of the year we ceased making any more tide-pole readings – enough information had been gathered – so I took the tide pole down and dismantled the long wave recorder, which was housed in a small hut at the end of the jetty. It was packed in a box and returned to England. The long wave recorder didn't involve us with much work – just changing the charts on the recording drum every week.

With the arrival of summer, I along with others looked forward to getting out and about on the island. We looked forward to exploring, helping Nigel (the biologist) in his work or going on a trip on a whale catcher. It had been the practice in the past to encourage these trips as it gave everyone a break and kept up the morale. Even when our workmates went off and we had to work harder, we knew that our turns would eventually come around. The Colonial Secretary, Denton-Thompson, was in full agreement with people getting away for a while.

So when at the beginning of November Nigel asked if I would go on a five-day trip on the sealing boat *Albatross* with him, to make a seal count all around the island, I leaped at the chance. Danny, Dave and Mike all agreed to cover for me. Pesca's manager said it was OK for me to go on one of the boats, but Captain Coleman said no. He said that I was employed to do meteorological work on the point and not to go gallivanting around on the island with Nigel. If Nigel wanted anyone's assistance on his work, the Captain said that he would have to employ someone from the sealing or whaling community. Eric also asked for permission to go out on a whale catcher for a few days; however, again the answer was no.

After our two requests to go on a trip, a directive from Captain Coleman appeared on the noticeboard in Discovery House which stated, 'No Whaling trips until the whole Point has been painted.' Everyone was up in arms, saying that they would not lift another paintbrush – especially as a painter and decorator were due to arrive in January. Also they said that Coleman seemed to forever be going on trips around to Husvik and Leith Whaling Stations socialising. It might have been part of his work to keep in contact with various companies, but his travels didn't go down well with everyone else confined to the point.

Although I was unable to go with Nigel on his seal survey on board the *Albatross*, I did help him to tag elephant-seal pups near the point. On our first trip around the bay, near the gun hut, it was 'Oh dear – we've left the tags behind.' So instead we just recorded the temperatures of seals. We crept up to sleeping or dozing seals, took their temperature and then crept away without disturbing them. We had to keep a wary eye on the big bull elephant seals.

The next time we remembered the tags and tagged seventy-five pups in the area near Penguin River. We clipped the tags into the loose skin under the pups' left flipper with the aid of a special tool which was like a pair of pliers. After the effort, I ended up with blistered hands. Two days later I tagged a further twenty-five pups. We tagged the elephant seals so we could track how long they lived if sealers shot them for their blubber and to find out how old they were before they were big enough to be killed.

Bird Island is off the north-west tip of South Georgia, and is heaven for any ornithologist or photographer. On the island there is a great abundance of birds – black-browed, grey-headed and wandering albatrosses, giant petrels, skuas and macaroni and gentoo penguins all nest on the island along with other seabirds.

All the birds are so tame you can walk right up to them while they are sitting on their nest to photograph them. There is now also a thriving fur-seal population and elephant seals. The fur seals of South Georgia were thought to be almost extinct after their slaughter in the 1700s and early 1800s, but they are making a great comeback now on Bird Island.

Early each summer the biologist spent a period of time on Bird Island studying the fur seals, making a census of the population and tagging many pups. To help him in this work, and so that he wasn't alone on the island, it was the policy for another member of King Edward Point staff to accompany and help him. So in November Nigel turned to me: would I accompany him? You can guess my answer, but of course Coleman said no. Coleman said that if he wanted help and company, he could pay someone from the whaling company to go with him. So Nigel told Coleman if I couldn't go, he would take his wife Jennifer with him, which is what he did. Coleman relented a bit and said I could go on the trip to Bird Island with them, but I had to come back straight away on the same day.

So we arranged it all, and I was going to move into the Bonners' bungalow while they were away to tend their plants and feed their chickens. The day for the trip arrived, and we set off from the point's jetty at 3.45 a.m. on board the *Albatross*. The sea was fairly smooth along the north-eastern coast, but cloudy with occasional snow flurries on the way up. On arrival at Jordan Cover, Bird Island, we anchored, and then we were taken ashore with supplies by scow. The way the sealers handled the scow was amazing, in and out of the kelp and timing our landing with the swell so we didn't get our feet wet.

While the Bonners and the sealers were sorting out the supplies and hut, Nigel told me to go off and see as much as I could before I had to leave – which I did, getting through a reel of film in the process. Then all too soon it was time to leave – how I wished I could stay!

On the way back, there were many icebergs scattered along the coast, and we had to weave in and out of them keeping close to the shore. The captain of the *Albatross* was Hauger, the top sealing skipper. I think he knew every bay and rock around the coast; everyone spoke with great admiration of his boating skills.

We passed the *Shackleton* in the Bay of Isles doing a hydrographical survey. Then we called in at Leith to collect some mail, and finally arrived back at the point about 9 p.m.

The next ten to twelve days, I looked after the Bonners' bungalow, and when it was time for them to come back Michael Meade went up for the trip and a chance to see the island.

We were expecting a visit from the Governor in January or February, but not one from the Colonial Secretary that summer. However, I think the news of discontent got back to him in the Falklands, as at the beginning of December he arrived at King Edward Point on board HMS *Protector*. At the subsequent cocktail party for him at the Colemans', he cornered me and asked me to take him for a walk around the bay, so he could do some filming. He knew from our talk in Port Stanley, when he interviewed me for my job, of my interest in wildlife and photography. The next afternoon I took the Colonial Secretary on a three-and-a-half-hour walk via Pesca Whaling Station, then the whalers' cemetery, where Sir Ernest Shackleton is buried, on around past Gun Point to Penguin River. There he took a cine film of king penguins, teals and elephant seals amongst other things. He said that he was very pleased with the walk as he spent all his time in Port Stanley sitting on his backside at a desk. I think he managed very well on the walk, seeing as he was missing one arm. He asked me to consider coming back to South Georgia once my contract had ended, as he thought I was doing a good job and had the right interests for it. Of course we talked about Captain Coleman and trips from the point, and he agreed that those trips should be encouraged.

Within a few days of our walk, Captain Coleman said that as long as the work was not jeopardised, we could go on trips from the point and not have to do the painting. The Colonial Secretary's popularity soared, and Captain Coleman learned that he was not running a ship now but a small outpost of the empire.

For Christmas, Eric and I decided to make a special menu for everyone in Discovery House. Early in December we made many copies of a photograph that Eric took of Pesca's church in the snow. We mounted these on printed sheets with the Mowatts' Christmas dinner menu. Ringdal, Pesca's manager, saw the photograph and asked if we could make some copies for him, which we duly did. On Christmas Day I had been on night duties, but Mr Ruddy awoke me at

10.30 with a cup of tea in bed – a rare privilege. I then got up and went to Discovery House to help Nancy and George prepare the dinner, which we served at 2 p.m. Ron Carter and I acted as waiters, and we spent the afternoon entertaining all the families from the point, and in the evening many guests from Pesca came and joined in. A lot of the parties ended in the Mowatts' bungalow at 3 a.m. the next morning, but I left early as I was on night duty again.

On New Year's Eve there was a party in the table-tennis room, and I acted as the barman for a while. Then at midnight I went with Eric, first-footing the Bonners. Later that night, Harry and one of his mates came visiting the point and were given drinks. This was fatal: they got into a fight with Bonski, a visiting whaling inspector, and in the process smashed up some sports equipment in the table-tennis room. At the subsequent trial for the offence, they were both sentenced to be imprisoned till the end of the whaling season, when they were deported back to Norway.

The only other time our police force was brought into action was when fighting broke out at Leith Whaling Station between two groups of 'Teddy boys' using flick knives. A whaler called for Jock and Basil's assistance (bring your truncheons), but by the time they arrived all had quietened down. However, they did make arrests and charged the offenders. At their trial, a few days later, which Captain Coleman presided over, three were fined and the others involved got off scot-free due to the lack of evidence and witnesses not coming forth.

During January, RRS *John Biscoe* on the way to relieve Halley Bay called in at the point and I renewed my acquaintance with Pat Canning, Chief Meteorological Officer for the Falkland Islands and the Antarctic. He looked cold, and it was the middle of summer, so I wonder what he felt like by the time he got to Halley Bay – the furthest-south FIDS base, at the base of the Weddell Sea. But it's surprising what a few tots of whisky can do. In fact, in January we had visits from all the FIDS Antarctic ships – *Shackleton*, *John Biscoe*, *Kista Dan*, and the Royal Navy ice protection vessel HMS *Protector*.

On 10 January a Russian tanker visited Pesca, bringing fuel oil for the whale catchers and to collect whale oil. Some of the crew took one of the tanker's lifeboats on a trip around to Cumberland West Bay to see the penguins and glaciers. On their way back they ran out of fuel and ended up drifting right across to the far side of Cumberland

East Bay. Nobody knew of their predicament until the evening, when we spotted red flares across the bay. We immediately dispatched the *Albatross* to investigate and eventually rescue the Russians and their lifeboat. It was a good job they landed where they did; otherwise they could have drifted right out to sea, and nobody would have known where they were.

Also early in January, Alan Best, Curator of Vancouver Zoo, arrived at the point on board RRS *Shackleton* – he had come to collect penguins for his zoo. While I was away on my whaling trip, Alan, with the help of Hauger and the *Albatross* had collected forty king, macaroni and chinstrap penguins. They were housed in various cages near the meteorological office. Many people assisted Alan to catch fish to feed his captives. I came along with my experience from the year before and constructed a fish trap and took it out by dinghy to set in the bay not far offshore amongst the kelp. One day Alan and I lifted the fish trap and found sixty fish inside – enough for two or three penguins' feeds. We had a bit of difficulty lifting the trap as it was all tangled up with the kelp. At the beginning of February the *John Biscoe*, on her return from Halley Bay, arrived with eighteen emperor penguins. We loaded the penguins we had at the point on board, and Alan sailed for Montevideo, from where they would be directly flown to Vancouver. Many of the whaling community at Pesca came to see him off and to see the emperor penguins as emperor penguins could not normally be found as far north as South Georgia.

I visited the plan at the whaling station many times to watch the process and to see the different types of whale (fin, sei and sperm). The local whaling inspector at Pesca took me around the bone locker and the rest of the factory, to see it all in action. It was an interesting experience.

Everything needed was catered for on the whaling station. They could be totally self-sufficient, having amongst other things a blacksmith (to straighten and reshape harpoons), carpenter, hospital, radio station (to keep in contact with their whale catchers), and shop (slop chest, run by Mareceleno Jensen). There was a cinema (*kino*), which showed three different films a week in the summer.

During the summer, I got to know some of the whaling personnel very well. For example, Eric and I got to know Arthur Brymer (South African doctor) and Ernest Gourker (hospital nurse/attendant) at

Pesca Hospital, and we spent several evenings visiting them for coffee in the hospital after going to the cinema.

One day Ernest turned up at the point with a small sack full of sperm-whale teeth and bones from the whale's ear for Eric and me – they were much sought after both by whaling workers and by people from the point. The whalers sought after the sperm-whale teeth to carve as they were made from solid ivory, and the bone from the whale's ear as it is shaped like a human head. I thought the ear bone looked like the head of Nikita Khrushchev, who was then the leader of Russia.

That summer (1959/60) the three whaling companies between them caught over 2,300 whales, of which 1,200 were fin whales. Only nine blue whales were caught. In the 1910s and 1920s the blue whale was the whale most hunted, being the largest of the whale family, reaching nearly 100 feet in length and weighing approximately 175 tons.

It is estimated that in that time the blue-whale population was reduced from 100,000 to 1,000. The factories preferred to have similar types of whales in at the same time to process, as oil from fin and sei whales had to be kept separate from sperm-whale oil. In the early days, in the bone locker, baleen plates from krill-eating whales were kept separate for the clothing industry, where they were used as whalebone to support ladies' corsets.

During the whaling season, whale meat was on the menu. The whalers' butcher left fresh lumps of meat to hang in wire cages (to keep the birds off) until they were black all over; then they would cut off the outer layer of meat, and then the rich red meat underneath would be ready for the pot.

Once, when the *R1* (one of Pesca's whale catchers) came in from hunting, I visited Rolfe Olsen, who showed me all around the whale catcher and introduced me to the gunner, Otto Larsen. Rolfe used to come over and visit Eric and me at the point in the winter. He was always asking us, "When are you coming out with us on a catcher?" This was in Captain Coleman's hands. Then, on 12 January, when Eric and I visited Pesca in the morning, when both the *R1* and the *R4* were in, Rolfe told me that the electrician was sick on the *R1*, so there was a spare bunk on the whale catcher. I saw the gunner, Otto Larsen, and he said that I could go out with them that afternoon. Mr Ringdal, the manager of Pesca, gave his approval, as did Danny, Michael and Dave in the meteorological office, and Captain Coleman. So I packed a few things and went on board the *R1* in the afternoon. Eric had similarly found space for himself on the *R4*.

Pesca operated six whale catchers (*RI–R5* and the *C. A. Larsen*), plus two other boats (the *Foca* and the *Petrel*) to help bring in the catch. In the summer, alongside the key at Pesca was the *Calpian Star* (originally the *Highland Chieftain*), which was used each year to transport whaling personnel between Norway and Grytviken.

There was a great rivalry between whale catchers, as crews got a bonus for each whale caught and for the amount of oil they produced from their catch. So all crews wanted to be with the top gunner. There was a restriction on the killing of undersized whales or females in milk, so whenever they spotted a family they allowed them to swim free; they only chased individuals or schools of whales.

When catchers are in there is always a film put on for them before they sail, so Eric and I joined the crew at the kino, then we both sailed at 6 p.m. The weather in the bay was calm and sunny, but the sea was very rough outside. We sailed around the south of the island, and then headed south-westwards. My first meal on board was fish balls, and seeing them sliding all over the plate in heavy seas rather quelled my hunger for them.

On 13 January, the weather started improving. I heard that the *R4* had killed a sperm whale off the south of the island in the early morning. So we went back accompanied by the *R3* to look for the rest for the pack, but had no luck. Then we headed south-west to where they caught fin whales. We sighted two fin whales at 6 p.m., and we chased them until 8 p.m., when we killed the first. The whale was then inflated with compressed air to make it float, and they stuck a radar target and flag into the body and cut the *R1*'s identifications mark into the tail. Then the whaler cast it afloat while we chased another, which we killed at around 8.30 p.m. As by then it was getting dark, we took both whales in tow till we rendezvoused with the tow boat *Foca*. They transferred our catch, and we sailed westwards. I spent over eight hours on the open bridge helping to look for the telltale signs of whales blowing.

On 14 January a gale blew up, so we changed course and headed for more sheltered and calmer seas off the eastern side of South Georgia to search for sperm whales. We saw no whales. During the day, the *R2* cut across our bows, nearly causing a collision. Many people waved their fists, and there was a lot of blue air from the gunner directed at the other boat. We played cards in the mess in the evening.

On 15 January the weather was cloudy, foggy and wet. We headed west again, but turned back in the afternoon, ending up near Bird Island, where we chased and killed a fin whale in the evening. The gunner gave me freedom of the whale catcher – the only restriction, for my own safety, was when they were likely to fire the harpoon. Also, when the winches were busy hauling in a whale I had to keep clear of the walkway from the bridge down to the gun platform.

On the 16th the day started off fine and sunny. We chased two fin whales at 6 a.m., but by 9 a.m. we had lost them in a bank of fog. We cleared the fog in the afternoon off the north-east coast of South Georgia, and then we travelled all down the east coast of the island, about thirty miles offshore. We didn't see a whale until we were at the bottom of the island. Then it turned out to be a family of fin whales with a youngster, so we left them alone. That afternoon I took over as helmsman. The weather was rough in the early afternoon, but it moderated in the evening.

On the 17th the weather was cloudy, but calm. We were 100 miles south of the island when we shot two fin whales in the morning and another in the afternoon. We chased several others, but couldn't catch them – they can swim nearly as fast as the whale catcher could steam; also, they can stay underwater for five minutes, which makes it difficult to guess where they are going to come up next. The boat just had to keep circling, and we hoped we were not too far away from where the whale would surface. At one time a whale swam under a lot of brash ice from where a berg had broken up, and we charged through the ice at top speed with the ice banging off the sides of the whale catcher. My mind flashed back to the *Shackleton*'s accident in 1957, when it was holed in ice and nearly sank, but we came through it safely and continued the chase.

The 18th was a rough and windy day with poor visibility. We chased two fin whales until we saw that they were mother and young, so then we left them alone. Otherwise we spent the day reading and playing cards.

The 19th was another rough day. We chased two whales for four hours in the afternoon, but we couldn't catch up with them while travelling at 16–17 knots. At 7 p.m. we sighted a school of twenty to thirty whales. The first ones we chased turned out to be a mother, father and young, so we left them alone. We then chased a single whale, which we shot just before it got too dark.

On the 20th it was a fine morning, but there were snow showers in the afternoon. However, the sea was not too rough until the evening, when we were 200 miles south-west of the island. We only saw whales in a family, which was left alone. In the evening we went fishing over the side of the whale catcher, with no bait, just shiny hooks, like mackerel fishing. In half an hour we had a tin bath full, so it was fresh fish for supper. We played cards for a good part of the day in the mess.

The 21st was quite a nice but cold day, with a lovely sunset in the evening. We didn't see any whales until about 6 p.m. The *R4* had got three whales when we rendezvoused with them, so we took over the chase while the *R4* took her catch in tow. We chased one whale until 10.30 p.m., but we never caught it even though we fired four harpoons at it.

On the 22nd there were gale-force winds, so we returned to Pesca. We came in along the south and east coast – we were very close in for shelter. The weather was quite nice in the lee of the land, and I saw many penguin rookeries along the shore. However, the weather turned foggy again before we got to the entrance of Cumberland Bay, and we had to cut our speed right down as we could see only about twenty yards (18 m) ahead. In the end we navigated in by radar. In the bay the weather was sunny and calm, and we tied up to the jetty at 6.30 p.m. I thanked Otto Larsen and Rolfe for the trip and made my way back to the point for a hot bath and a change of clothes. I then went off to Eric's to compare our adventures. The period we were out was not very successful. We must have been a jinx on them as just over a week later Pesca had its biggest catch in one day. Also, in February the *R1*, on occasions, was catching four to six whales a day.

On the last day of January I sailed on the *Albatross* to Leith for the South Georgia Sports Day. We left the point at 6.30 a.m. and returned at 8.30 p.m. The day was sunny, but very windy. I didn't go for the sports so much as to see both Leith and Stromness Whaling Stations. At Leith I met last season's *Southern Gem* gunner, who visited us at Signy. He was a plan foreman now at Leith. We had a long chat, and he told me all about his work and about our penguins which went to Edinburgh Zoo. He also asked after the Signy crowd. While I was there they brought ten sperm whales in. This was the day that Pesca had their biggest catch of the season (thirty-nine).

In February, Mr Boris (architect) visited the island to search for a site on King Edward Point to build a new three-storey building to be called Shackleton House. It was to be built up past the Bonners' house on a flat piece of ground at the head of the point, up towards Shackleton's Cross. He wanted a detailed contour map of the point linking in the distances between all the buildings. Also, he wanted the map to show the height of the dam on the side of Mount Duce, which was our source of water for all the buildings, and the distance of the side of the mountain from the flatter areas. As there was not such a finely detailed map available, and no surveyor of the island (the cheek of it!), I volunteered to produce a map for him with the help of the two prisoners. We borrowed all the equipment to do the job from Leith Whaling Station – for example, a surveyor's theodolite and tripod, a steel tape measure and surveying poles. The only surveying experience I had was assisting Doug Bridger on Signy, but I had a fairly good knowledge of geometry and trigonometry.

I started work at the benchmark on the side of the customs shed by the jetty, and took a straight line up to the Bonners' house. To assist me, I had the two prisoners (Harry and his mate from the New Year's Eve incident) helping to hold the poles, to mark the positions on the ground every 100 feet (30 m), and to run out the tape. All three of us got on well, and I think they enjoyed themselves. It was much better for them than being given menial tasks, such as the cleaning-up. But each night they were handed over to one of the constables to be locked up, while I was busy with the slide rule and tables working out each day's figures. All this I did between my normal meteorology duties until Captain Coleman asked the other two meteorologists to take over some of my work. In the end, the architect seemed very satisfied with what I had produced and gave me £20 for my efforts. Later Shackleton House was built – and to think its position was all based on my survey work!

The 1st of March saw the long-awaited visit from the Governor of the Falkland Islands and Dependencies on HMS *Protector*. The Colemans put on a cocktail party for him. He cornered me later, and we had quite a long chat about my whaling trip, Signy Island, and the new Falkland Island bird stamps. These stamps, in the first year of their issue, were supposed to produce an extra £20,000–30,000 of revenue for the Falklands.

Off HMS *Protector*, I received a letter which had only taken two weeks from England – this must be a record! The Governor departed the next day for Leith Whaling Station.

In March a Mr Greyber from Germany arrived at Pesca to collect seals and penguins to sell to European zoos. We didn't have anything to do with him and his activities as he stayed at the whaling station all the time. But I do remember seeing six poor elephant seals all crated up and ready to be shipped out on a whaling boat at the end of the season. I do know he had one rare capture for South Georgia – a crab-eater seal, which he caught on the beach near Pesca's radio station. I suppose Captain Coleman had to give an export licence for both lots of penguins and seals. Alan Best was a caring man, whereas Mr Greyber was purely a businessman. I don't think he had much respect for his cargo, but rather he just thought about the money he could make out of it.

Mr Greyber left South Georgia on the *Calpian Star* with his ninety penguins and eighteen seals destined for various European zoos. However, on her way to Husvik the *Calpian Star* hit the bottom and bent her rudder. The Captain was relieved of his post and Bordal, gunner of the *R2*, took over the boat's captaincy. The *Calpian Star* had to go for a dry-dock repair, but where? The dry dock at Stromness was not big enough for such a large boat. In the meantime, two *R* catchers had to tow her wherever she went. They requested an ocean-going tug – a special type of boat – to tow her. Then they moored her in Jason Harbour, Cumberland West Bay, till it was possible to tow her to a dry dock for repair. One day, the *Calpian Star* started to drag her anchor, and Ringdal, who was on board, tried to contact Pesca by radio to get assistance from the boats there. But the winter manager, Rogner, was drunk and wouldn't get off the radio and kept interrupting their broadcast. Ringdal came ashore by motor boat, gave Rogner the sack on the spot and installed Torr Torson in his place as winter manager. No tug would be available until late April, so the *Calpian Star* had to stay moored in South Georgia waters with two *R* boats in attendance until the tug *Atlantic* came. I saw her still there when I left on RRS

John Biscoe on 14 April. I wonder what happened to the cargo of penguins and seals destined for European zoos.

On 22 March, it was a day of gales and very strong gusts of wind. A trawler which had been fishing in the South Orkney area came into Grytviken for repairs. She had once landed on Powell Island, where they had made a couple of ringed-bird recoveries – one skua and one giant petrel. It turned out that the giant petrel was one I had ringed on Signy Island.

All throughout the summer I continued to make regular balloon ascents every two to three days, when the weather permitted. Often the balloon would rise almost vertically until it had reached 6,000–7,000 feet (1,828–2,134 m). Then, after it had got out of the shelter of the high ground, the balloon would suddenly shoot off in winds of 50–60 knots, and the handles on the theodolite would be turning like mad to keep up with the disappearing balloon.

Due to the mountain range, at times we used to get some very interesting cloud formations. We had some cloud formations swirling around all the time, and others were stationary and saucer-shaped, just like flying saucers or stacks of plates.

During the whaling season our radio staff were supposed to help us with the reception of weather information from various countries, but they were often too busy and didn't bother, or were drunk, or said the broadcast was unreadable when, in fact, Danny and I could hear them clearly; so we took the information ourselves and didn't rely on them for help.

For entertainment we sometimes went to the *kino* (cinema) at Pesca, which put on three different films a week during the summer. Anyone from the point could go – we just had to buy a season ticket from the slop chest. If whale catchers were in, there was a matinee show; otherwise there was an evening performance. The films were a mixture of British, American, German and French films, all with Norwegian subtitles. We were way ahead of most of the audience when the programme was in English, but completely lost when the dialogue was in French or German. Before each film show a motor boat used to come to the point to take any viewers across to the *kino*. The boat had a very powerful engine which used to tick over very slowly with a plop, often blowing lovely smoke rings from its exhaust straight up into the sky. The main use of this boat was for pushing and towing whales back to the plan, so the deck was low, flat and completely open. There were no deck handrails. If we went for the ride we had to stand on the open deck and hope we didn't fall over the side. The boat was very stable and didn't roll or pitch.

Eric and I, most times, walked around the edge of the bay, as usually we went visiting before or after the show. I remember visiting the hospital with Eric once to have coffee with Arthur and Ernest. At the time, there were only three patients in the hospital, for broken bones and other injuries. Two were guitarists and were playing away, having a good old time, and they were joined later by three other guitarists, who were members of the *Calpian Star*'s crew. The hospital was really rocking and resounding to the sound of music that night.

Also, for entertainment we organised a crib match between the point and the whaling station. We had a team of six players (Ron Carter, Ron Houlton, Danny, John, Jimmy and me), and we each played six games. We held the home match at Discovery House. I lost my match 4–2, and our team lost overall. On the return match, held in the lounge of the *Calpian Star*, I drew my match 3–3, but our team still lost overall again.

We used to sometimes have evenings at the Mowatts', playing cards (Newmarket) or Monopoly. Usually the group consisted of John, Ron Carter, Bill (the painter), the Mowatts and me.

We also spent quite a few evenings doing our own film showings. Usually it was Nigel (with his films of Bird Island or of various sealing trips) or Danny (these showed the point over the years since he first came to South Georgia) or me (with my slides of Signy Island).

Most of the summer when I wasn't working, and the weather wasn't too inclement, I went walking and scrambling around the local area. Early in the summer I was restricted to the shoreline due to slushy snow cover, but as the summer progressed the snow receded until there was little on the ground below 1,000 feet (305 m). A lot of the time I was accompanied on trips by Eric.

One of my most regular scrambles was from the point straight up the side of Mount Duce via the dam which held our water supply. From the top there was a lovely view all over Cumberland East Bay. At other times I went around the coastal path past Pesca, Gun Point and Penguin River to Hestesletten Plain, where I had to step over or around many elephant seals. By mid-November I had found the first gulls' nest with eggs along the cliffs by Shackleton's Cross. I also found the terns had started to lay on the flat stony ground.

Just before Christmas, Eric and I made our first visit to Maiviken. We took the track around the bay to Pesca. We headed up past Grytviken Church to Dead Man's Pass, then down the valley between Spencer Peak and Mount Hodges. The scenery descending to the bay overlooking Cumberland West Bay was very pretty, with streams running down via several lakes and waterfalls to the sea. The flowering mosses covered the ground in places. Mrs Colman heard us talking about our trip and asked us to take her, which we did a week later. Unfortunately, on this occasion the weather wasn't so kind, and we had the odd snow and rain showers. Even so, she was well pleased with the trip.

On another trip to Maiviken, while scrambling along the top of the scree slope on the Spencer Peak side of the valley I found two different kinds of ferns, which I took back for Nigel.

In January I took Dave to Maiviken, and we found a cave. Inside there was some tinned food, a fishing line, pots and pans, and a cup. Most likely it had been left there by the sealers to use when sealing in the bay. On the walls of the cave many people had carved their names. One name was 'H. Bordal 1941' (now *R2*'s gunner). We added our names and the date.

I found another new fern while exploring in Maiviken Valley, and Nigel said the fern had never been recorded in South Georgia before. The next day I took Nigel and showed him where I found the fern. We then explored up near Dead Man's Pass and on either side of the valley, collecting specimens of all the ferns we could find on South Georgia.

We also collected moss specimens, land beetles, and water bugs from one of the small pools. All these specimens Nigel was going to send back to the British Museum.

One bit of fun I used to have was coming down from the top of the scree slope. As all the scree was at an unstable angle, any slight movement got the whole surface moving. I used to come down with a shower of sliding shale – as long as I kept moving and standing upright, I was all right.

One day, instead of going to Maiviken, when I got to Dead Man's Pass I then doubled back along the scree slope below Mount Hodges overlooking the whaling station and then came down by Gull Lake and Pesca's radio station. The view from above the whaling station over the bay was quite spectacular.

Eric and I made two trips to Cumberland West Bay via Echo Pass. The first trip was on New Year's Day. We left at 8 a.m. and returned at 7 p.m. – a good day's stroll! We went via Pesca up by Pesca's radio station to Gull Lake, between Mount Hodges and Brown Mountain, then we went around the back of Mount Hodges to Echo Pass. Here we were about 1,000 feet up (305 m), and we had to cross a snowfield before descending to the other side of Cumberland West Bay.

We first visited the gentoo rookery, where about half the eggs had hatched. Then we went along the coast to the face of the large and most active Neumeier Glacier.

The day was sunny and warm, and we ended up taking our shirts off and sunning ourselves sitting on rocks close to the face of the Lyell Glacier, eating our sandwiches. All the time there were rumbles and bangs from the glacier, and the occasional big splash as lumps of ice fell away from the glacier's face. There was a group of king penguins on the beach looking all smart, so we took a photograph. While exploring, we had to cross several streams, which meant boots and socks off and trousers rolled up to the knees. The water was certainly invigorating as it had come straight off the glacier!

We found a piece of volcanic rock, which we presented to Nigel on our return. It was a tiring but very enjoyable day and we both ended up rather sunburnt.

Hestesletten Plain (Norwegian for 'Horse Plain') was so named after people had released horses on to the plain in the early 1900s. There were no horses there now. We made many trips to Hestesletten Plain and further in past the lakes to the hanging Hamberg Glacier. In this area we found gulls, terns and skuas nesting. Along the cliffs sooty albatrosses were skimming back and forth, but we couldn't find any of their nesting sites. One time we climbed up a valley, and from the top of the ridge overlooking Moraine Fjord we looked down towards the Harker and Hamberg Glaciers. On 3 January I saw a young fur seal on the beach near Penguin River. I duly photographed him, and on my return, I told Nigel of my sighting. The next day Nigel and several members of the point staff went around to see the fur seal, and he had obligingly waited for them. We rarely saw fur seals in this part of the island.

Late in January, Dave, Ron Carter, and I walked to Penguin River, then we went up the gorge behind Brown Mountain towards Gull Lake. At one point, Ron slipped down a steep slope into a stream at the bottom, losing one boot and getting rather damp. While Dave stayed with Ron, I hurriedly retrieved Ron's boot about half a mile down the stream, just before the stream disappeared into a snow tunnel. Ron got back OK, but he was rather shaken, bruised and damp. It was the first time that Ron had been so far from the point, and he said never again!

In mid-March I walked to Discovery Point, where John Quigley met me with the motor boat, and we then went up Moraine Fjord and photographed both the Hamberg Glacier and the Harker Glacier.

Then we landed on the other side of Dartmouth Point and climbed up on to high ground overlooking the Nordenskjold Glacier; we saw several giant-petrel nests with young still in them.

By time it got to late March we had some very strong gales with gusts of over seventy knots. The sea surface was whipped up, and spray was blowing everywhere, which reduced visibility and coated all windows with a layer of white salt.

The whaling season ended on 31 March. Pesca was the station with the highest catch on the island. *R3*'s gunner was top with 233 whales; Otto Larsen of *R1*, whom I went out with, was second. At the end of the season, all the whale catchers were stripped down of whaling equipment for overhauling at the station during the winter, then refuelled for their journey home to Norway. However, some of the whalers had to wait behind while they were waiting for the *Calpian Star* to be relieved with an Atlantic tug.

These whaling crews became fed up just having to sit there and wait, when they could have been on their way home to Norway.

Before the season finished, however, a dentist, Dr Jacobie (ex-German Second World War pilot) visited Pesca and we all went over to have a check-up. He asked me when my teeth had last been checked. I said it was three years ago. He rubbed his hands and said, "I usually reckon on three filling per year." But he couldn't find any to fill – in fact, all my teeth were OK.

By the first week in April, winter was returning with frequent snow showers and everywhere was white again, just as it was when I arrived in June the year before.

After all Captain Coleman's fuss about getting the buildings on the point painted, Bill, the painter, with the help of the two prisoners had all the outside decorating done before the winter arrived.

Further south, in the Antarctic the ice conditions had been particularly bad along the Graham Land Peninsula and the FIDS ships had got stuck a few times and had to be helped by the American icebreaker USS *Glacier*. Base F had to be closed, and Base Y was relieved by air as they could not get the boats anywhere near the base. This made the timing of my departure from South Georgia very uncertain. In the end, RRS *John Biscoe*, after making her final visit to Signy, called for me on 14 April.

I said farewell to all my friends at the point and boarded RRS *John Biscoe*, meeting up once again with Fergus O'Gorman, Mike Rhodes, Jim Young, Robin Perry and many other Fids I had travelled down with way back in 1957.

The journey to the Falklands wasn't too rough, though it was a bit damp and foggy at first, but by the time we reached Port Stanley on 18 April the weather was fine and sunny, but a little cool.

Part Four: Port Stanley Again and the Journey Home

Port Stanley

Nothing was straightforward that summer. A new hut was on board the RRS *John Biscoe*, and they were supposed to offload it at Stonington Island, Base E, but due to bad ice conditions the *John Biscoe* could go nowhere near the base. They therefore decided to unload the hut in Port Stanley along with all the sand and shingle for making the foundations. The crew of the *John Biscoe* did not want to take the hut all the way back to England! The two builders, Jim Shirtcliffe and Harry Dollman, who were going to erect the hut on Stonington Island, were given the job of building the hut at the top of the hill beside the site of the old meteorological office. These were far-different building conditions from what Jim and Harry had expected to be working under! So, the first three to four days in Port Stanley were spent unloading and moving all the materials from the *John Biscoe* by tractor and trailer up the steep Philomel Hill to the new site.

Another change of plan: instead of going straight home to England, one of the aircraft on the Graham Land Peninsula had damaged a wing. A new wing had been flown to Montevideo, so now the *John Biscoe* was to make a special trip up to Montevideo to collect the wing, and then make a dash down to Deception Island to deliver the wing before the winter set in and the sea froze up. Then it would be back to Port Stanley before starting the journey home. I thought this would be a good opportunity for me to see a bit of Graham Land and Deception Island, which I had never seen. But they decided that only the crew would make the journey. Instead we (the Fids) had to stay in the Falklands while the *John Biscoe* made her extra journey.

On my second day in Port Stanley, the Colonial Secretary, Denton-Thompson called me into his office at the secretariat; he tried to persuade me to come back to South Georgia. I explained that I was on contract to the Met Office, who had only loaned me to the FIDS for three years – to come back I'd have to resign from the Met Office and I was not prepared to do so at that time. As I was the first person out of South Georgia, via the Falklands, since his visit, we then had a talk about the point and Captain Coleman. I told him that everyone was a lot happier since his visit and that Captain Coleman had relented and was letting people go on trips away from the point. After about three-quarters of an hour, I left his office and visited the hospital to see the Matron and staff that had looked after me the year before.

My friends Eric and Eileen again said that I could stay with them, while the other Fids stayed at various lodgings, including The Ship Hotel. So now we were back to the usual round of partying around the town – at Peter Hale's, Clem and Sadie's, Mary Woods's, and the homes of several other friends. Much of the time I was accompanied by Kate Atkinson, who was a school teacher from England on a three-year contract to the Falkland Island Government. Kate later married Brian Waudby, one of the meteorological-office forecasters. When not partying, many evenings were spent at either Eric and Eileen's or Kate's flat playing records or talking, often along with Jim Stammers, Mary Woods, Robin Perry and Eirene Duff. The three girls all had flats in a house on Ross Road overlooking the harbour. We also went to the usual Sunday-evening film show at the Town Hall and to the odd dance or two. I saw Evie on quite a few occasions as she used to babysit for Eric and Eileen. She had given up her job at the hospital and now worked in the Falkland Islands Store.

One evening, Glyn Pugh and Brian Waudby, from the meteorological office, Jim Shirtcliffe and I all went out drinking at the Glue Pot, a local watering hole. When we left, not too steadily, Jim said he'd give me a lift back to Eric's on his motorbike, but we couldn't find it! I think it was a good job too as it was a big and powerful motorbike and we'd all had too much to drink.

I went on many walks, often at the weekend accompanied by Kate, to various local places of interest, such as Sappers Hill, The Two Sisters, York Bay and Eliza Cove. Twice Jim Stammers and I went collecting freshwater specimens – once from the Mile Pond and the other time near the filtration plant in Mullet Creek. Jim reckoned he had found three previously unrecorded creatures in the Falklands.

On 10 May we planned to go by the government boat *Alert* to Kidney Island to see the wildlife there (sea lions and penguins).

The day before, I was passing the Colonial Secretary's office when the window flew open and Denton-Thompson shouted, "I say, Richards – are you going on the Kidney Island trip tomorrow?"

"Yes, sir," I replied.

"Good. I'll see you then." Back came his reply.

However, the next day the weather was bad with rain and strong winds, so they cancelled the trip.

The Falkland Islands to England

The 11th of May saw the *John Biscoe*'s return after her trip to deliver the aircraft's wing to Deception Island. So all the Fids bade farewell to the crowd that we had made friends with in Port Stanley and boarded her again. The next day we sailed at 9 a.m. To see us off were the usual FIDS-office and meteorological-office workers.

Bill Johnson, commonly known to the Fids as Captain Kelley, captained RRS *John Biscoe*. The first officer was Tom Woodfield, who had been the second officer on RRS *Shackleton* in 1957. The regime on the *John Biscoe* seemed a bit more rigid than on the *Shackleton*, and Captain Kelley said he only wanted professional meteorologists making weather observations on his boat. So as Robin Perry and I were the only meteorologists by profession, we ended up doing meteorology duties on alternate days. Unlike on the *Shackleton*, we didn't have to throw a canvas bucket over the side to collect a sample of seawater and put a thermometer into it to find the sea temperature as there was a recording instrument on the bridge for it. On the return trip, as on the downward journey, we had a captain's inspection every Sunday morning. Then the Captain along with several officers inspected all cabins, FID'ary and galley – this meant a hurried tidy-up and no lingering in the bunks.

The journey from the Falklands to Montevideo was slow, averaging only eight knots. The seas were rough with strong winds, and there was fog and rain on the first three days. It did brighten up the day before we got to Uruguay, but we still had heavy swell and strong winds. The journey took up five and a half days. We only spent one day in Montevideo on the way home, so Jim Stammers and I made the most of it, sightseeing around town, buying souvenirs, taking photographs and treating ourselves to a big steak meal at the Victoria Plaza Hotel. The steak was so large that neither of us could manage all of it!

The next day, 18 May, we left Montevideo at 11 a.m. for Southampton. With a following sea and wind we were now making eleven knots. On the way down the River Plate Estuary we passed the *Calpian Star* being towed in by tug *Atlantic*. Apparently, there were four aircraft standing by at Montevideo to fly all the 240 whalers back to Norway. I don't know what the fate of Mr Greyber and his collection of penguins and seals was.

For the next four days we travelled up along the Brazilian coast, all the time with favourable winds and seas. We were now averaging twelve knots. Several land birds were blown out to sea and landed on the decks, which had Fergus duly trying to identify them. We had occasional schools of porpoises swimming and porpoising around the bows of the ship. The weather was rather cloudy and humid, but it got warmer day by day until after six days the sea and air temperatures were both up to 27°C (80°F).

Two days later we passed three islands belonging to Brazil. On one the Americans had a tracking station for their rockets, and on one of the others was a Brazilian prison settlement. We were now only three degrees south of the equator, which meant that the weather was very hot and sticky. We spent quite a lot of the time hanging over the side of the boat watching gannets diving and chasing flying fish.

At midday on the 28th we crossed the equator, and two days later we entered the north-east trade-wind belt, which reduced our speed down to ten knots. We saw many Portuguese men-of-war jellyfish and at a distance one whale. We did several lifeboat drills, and we spent several mornings holystoning the decks.

On 1 June we passed Cape Verde, and a day later the sun was directly overhead. On 3 June we crossed the tropic of Cancer. Then we were out of the tropics, the temperature dropped slightly, being 24°C (76°F) during the day.

On the 6th at breakfast time we passed Madeira, where orographic clouds covered the tops of the mountains. Otherwise the day was warm and sunny. By the 9th we were off Cape Finisterre in the morning. The sea was very busy with shipping – at one time we could see twelve different ships. The day was cloudy with a heavy swell, but little wind. Now for the Bay of Biscay!

We were soon through the bay, and we sighted the Brest Peninsula on the evening of the 10th. There was still plenty of shipping about.

The weather was now sunny, but cool. I made my last weather observation.

The next day at about noon we saw Portland Bill – our first sight of England for nearly three years. The weather had now deteriorated, with low cloud and sea-fog patches. We followed the coast along towards the Isle of Wight and anchored off the Needles at 3.45 p.m. A pilot boat came out to see us in, but Captain Kelley said, "We're staying here until tomorrow." They had given the following day as our estimated time of arrival, and consequently they had made all arrangements for our arrival for the next day too, so that evening we had one final farewell party in the FID'ary.

The next day we upped-anchor at 11.30 a.m., when the pilot came on board. The weather was blowing a gale with heavy rain. We came in past the Saunders-Roe hangars, where the two Princess flying boats were housed, then passed the Foley Oil Refinery to come alongside Southampton Berth 37 at 2 p.m. – just as it stopped raining. I finally cleared customs at 3.30 p.m. We were all welcomed home by the Governor of the Falkland Islands (who was on home leave in England), Sir Vivian Fuchs (Bunny), Captain Turnbull, Tom Flack and several officers of RRS *Shackleton*. Also there was George White, who was with us on Signy and had come back on the *Shackleton*, plus, of course, my parents.

So ended the journeying and adventures. Now for a good holiday!